STATISTICS

WITH MICROSOFT® EXCEL

FIFTH EDITION

Beverly J. Dretzke

University of Minnesota

PEARSON

Boston Columbus Indianapolis New York San Francisco Upper Saddle River
Amsterdam Cape Town Dubai London Madrid Milan Munich Paris Montreal Toronto
Delhi Mexico City Sao Paulo Sydney Hong Kong Seoul Singapore Taipei Tokyo

Library of Congress Cataloging-in-Publication Data

Dretzke, Beverly Jean.

     Statistics with Microsoft Excel / Beverly J. Dretzke. -- 4th ed.

        p. cm.

     Includes bibliographical references.

     ISBN 978-0-321-78337-0

     1. Statistics--Data processing. 2. Microsoft Excel (Computer file) I. Title.

QA276.45.M53D74   2012

519.50285'554--dc23

                                  2011031903

The author and publisher of this book have used their best efforts in preparing this book. These efforts include the development, research, and testing of the theories and programs to determine their effectiveness. The author and publisher make no warranty of any kind, expressed or implied, with regard to these programs or the documentation contained in this book. The author and publisher shall not be liable in any event for incidental or consequential damages in connection with, or arising out of, the furnishing, performance, or use of these programs.

Reproduced by Pearson from electronic files supplied by the author.

ISBN-13: 978-0-321-78337-0
ISBN-10: 0-321-78337-9

www.pearsonhighered.com

Preface

The fifth edition of Statistics with Microsoft® Excel was prepared using Office 2010. If you are a brand new Excel user, I have included many screen prints in this manual, so there should be no difficulties in learning how to carry out data analysis procedures.

Every chapter in the fifth edition of Statistics with Microsoft® Excel has been updated to include screen prints from the 2010 version of Microsoft® Excel. Additional updates are listed below.

- **Chapter 4: Frequency Distributions**. Additional instructions on how to use the FREQUENCY function and how to enter data arrays.
- **Chapter 5: Descriptive Statistics**. Additional instructions on how to weight responses to adjust for survey nonresponse.
- **Chapter 6: Probability Distributions**. Updated binomial, hypergeometric, Poisson, normal, F, t, and chi-square distribution functions
- **Chapter 7: Testing Hypotheses About One Sample Means**. Updated normal and t-distribution functions.
- **Chapter 10: Correlation**. A revised data set and an updated function for finding ranks.
- **Chapter 12: Cross tabulations**. Updated chi-square functions and revised chi-square data set.

All data sets used for analysis examples are intentionally small so that data entry time will be minimal. To make it even more convenient for users of this manual, many of the data sets are available on: www.pearsonhighered.com, as well as netfiles.umn.edu. Instructions for accessing these web sites are given in Chapter 1.

Although I do provide explanations and interpretations for many statistical analyses, this manual is not intended to be an introductory statistics textbook. Rather, it is a technology manual designed to accompany a text. Please refer to an introductory statistics textbook for a complete explanation of any of the analysis and procedures that are included in this manual.

I am very grateful to students, instructors, and other users for their feedback. Whenever possible, I have tried to incorporate revisions that address their comments and suggestions. I am also grateful to Paul Schaleger for his expert technical advice. Excellent editorial support was provided by the people at Pearson, most notably Sonia Ashraf and Joe Vetere.

If you have any comments or suggestions that you would like to make regarding this manual, please send me a message.

Beverly J. Dretzke
University of Minnesota
Twin Cities Campus
(dretz001@umn.edu)

Contents

Chapter 1: Getting Started

1.1. Introduction and Overview 1
    Versions of Excel 1
    Versions of Windows 2
1.2. What May Be Skipped 2
1.3. Excel Worksheet Basics 2
    Using the Mouse 2
    Starting Excel 3
    Exiting Excel 3
    Layout of Worksheets 3
1.4. Dialog Boxes 5
1.5. Accessing Excel Files on the Web site 5
1.6. Saving Information 6
1.7. Printing 6
1.8. Loading Excel's Analysis ToolPak 9

Chapter 2: Entering, Editing, and Recoding Information

2.1. Opening Documents 13
    Opening a Brand-New Worksheet 13
    Opening a File You Have Already Created 13
2.2. Entering Information 14
    Addresses 14
    Activating a Cell or Range of Cells 14
    Types of Information 15
    Filling Adjacent Cells 15
    Series 16
2.3. Editing Information 18
    Changing Information 18
    Moving and Copying Information 19
    Moving and Copying an Entire Worksheet 20
    Dragging and Dropping 21
    Inserting or Deleting Rows and Columns 21
    Changing the Column Width 22
2.4. Formatting Numbers 22
    Decimal Points 23
    Currency 24
    Percentage 24
2.5. Recoding 25
2.6. Sorting 28

Chapter 3: Formulas

3.1. Operators 31

Order of Operations 31
Writing Equations 32
Relative References 34
Absolute References 34
3.2. Using Formulas in Statistics 36
Sample Research Problem 36
Mean 36
Deviation Scores 40
Squared Deviation Scores..41
Variance 42
Standard Deviation...44
z-Scores 45

Chapter 4: Frequency Distributions

4.1. Frequency Distributions Using Pivot Table and Pivot Chart 49
Sample Research Problem 49
Frequency Distribution of a Quantitative Variable 49
Frequency Distribution of a Qualitative Variable 55
4.2. Frequency Distributions Using Data Analysis Tools 59
Sample Research Problem 59
Steps to Follow to Create a Frequency Distribution and a Histogram for
Grouped Data..59
Steps to Follow to Create a Frequency Distribution and a Histogram for
Ungrouped Data..68
4.3 Frequency Distributions Using the FREQUENCY Function 74

Chapter 5: Descriptive Statistics

5.1. Data Analysis Tools: Descriptive Statistics......................................77
Sample Research Problem 77
Steps to Follow to Obtain Descriptive Statistics for One Variable 77
Interpreting the Output 80
Steps to Follow to Obtain the Population Variance and Population Standard
Deviation 82
Steps to Follow to Obtain Descriptive Statistics for Two or More Variables 85
Missing Values 87
5.2. Functions: Descriptive Statistics..87
Sample Research Problem 88
Steps to Follow to Obtain Descriptive Statistics for One Variable 88
Steps to Follow to Obtain Descriptive Statistics for Two or More Variables 91
5.3. Pivot Table: Descriptive Statistics 92
Sample Research Problem 92
Steps to Follow to Set Up a Pivot Table 92
Changing or Removing Summary Measures from a Pivot Table...........97
5.4 Weighting to Adjust for Survey Nonresponse......................................99
Sample Research Problem 99
Steps to Follow to Weight Responses Based on One Variable...............99

Chapter 6: Probability Distributions

6.1. Discrete Probability Distributions 103
    Binomial Distribution 103
    Hypergeometric Distribution...108
    Poisson Distribution 111
6.2. Continuous Probability Distributions 115
    Normal Distribution 115
    t Distribution..122
    F Distribution..126
    Chi-Square Distribution...129

Chapter 7: Testing Hypotheses About One Sample Means

7.1. One-Sample z-Test...131
    Assumptions Underlying the z-Test 131
    Sample Research Problem 132
    Steps to Follow to Analyze the Sample Research Problem 132
    Interpreting the Output..137
    Confidence Interval for the One-Sample z-Test...139
    Interpreting the Confidence Interval 139
7.2. One-Sample t-Test...140
    Assumptions Underlying the t-Test 140
    Sample Research Problem 140
    Steps to Follow to Analyze the Sample Research Problem 140
    Interpreting the Output..151
    Confidence Interval for the One-Sample t-Test...152
    Interpreting the Confidence Interval 153

Chapter 8: Testing Hypotheses About the Difference Between Two Means

8.1. t-Test for Two Independent Samples 155
    Variances are Not Known and Are Assumed to Be Equal 155
    Assumptions Underlying the Independent Samples t-Test 156
    Sample Research Problem 156
    Steps to Follow to Analyze the Sample Research Problem 156
    Interpreting the Output..158
    Variances Are Not Known and Are Assumed to Be Unequal.............................159
    Assumptions Underlying the Independent Samples t^*-Test..............................160
    Sample Research Problem 160
    Steps to Follow to Analyze the Sample Research Problem 160
    Interpreting the Output..162
8.2. Paired-Samples t-Test...163
    Assumptions Underlying the Paired Samples t-Test...164
    Sample Research Problem 164
    Steps to Follow to Analyze the Sample Research Problem 164
    Interpreting the Output..166

8.3. *z*-Test for Two Independent Samples 167
    Assumptions Underlying the *z*-Test 167
    Sample Research Problem 168
    Steps to Follow to Analyze the Sample Research Problem 168
    Interpreting the Output 169

Chapter 9: Analysis of Variance

9.1. One-Way Between-Groups ANOVA 171
    F-Test 171
    Assumptions Undelying the *F*-Test 171
    Sample Research Problem 171
    Using Analysis Tools for One-Way Between-Groups ANOVA 172
    Interpreting the Output 173
9.2. One-Way Repeated Measures ANOVA 175
    F-Test 175
    Assumptions Underlying the *F*-Test 175
    Sample Research Problem 175
    Using Analysis Tools for One-Way Repeated Measures ANOVA 176
    Interpreting the Output 177
9.3. Two-Way Between-Groups ANOVA 180
    F-Test 180
    Assumptions Underlying the *F*-Test 180
    Sample Research Problem 180
    Using Analysis Tools for Two-Way Between-Groups ANOVA 181
    Interpreting the Output 182
9.4. *F*-Test for Two Sample Variances 185
    Assumptions Underlying the *F*-Test 186
    Sample Research Problem 186
    Using Analysis Tools for the *F*-Test for Two Sample Variances 186
    Interpreting the Output 188

Chapter 10: Correlation

10.1. Pearson Correlation Coefficient 189
    Sample Research Problem 189
    CORREL Function 190
    Correlation Matrices 192
    Interpreting the Output 194
    Scatterplot 194
    Modifying the Scatterplot 197
10.2. Spearman Rank Correlation Coefficient 199
    Sample Research Problem 199
    RANK.EQ Function and Correlation Analysis Tool 200
    Interpreting the Output 203

Chapter 11: Regression

11.1 Two-Variable Regression 205
    Sample Research Problem 205
    Steps to Follow to Analyze the Sample Research Problem 206
    Interpreting the Output..208
11.2. Multiple Regression 210
    Sample Research Problem 210
    Steps to Follow to Analyze the Sample Research Problem 211
    Interpreting the Output..213
11.3. Dummy Coding of Qualitative Variables...215
    Sample Research Problem with Two Groups 216
    Steps to Follow to Analyze the Sample Research Problem 216
    Interpreting the Output..218
    Sample Research Problem with Three Groups..220
    Steps to Follow to Analyze the Sample Research Problem 220
    Interpreting the Output..222
11.4. Curvilinear Regression 224
    Sample Research Problem 224
    Steps to Follow to Analyze the Sample Research Problem 224
    Interpreting the Output..226
    Scatterplot of a Curvilinear Relation..228

Chapter 12: Cross Tabulations

12.1. Cross Tabulations Using the Pivot Table 233
    Sample Research Problem 233
    Cross Tabulation of Two Qualitative Variables 233
    Cross Tabulation of a Qualitative and a Quantitative Variable......................237
    Cross Tabulation of Three Variables..242
    Missing Data 247
12.2. Chi-Square Test of Independence..249
    Assumptions Underlying the Chi-Square Test of Independence.....................249
    Sample Research Problem 249
    Using the Pivot Table for Observed Cell Frequencies 250
    Using Formulas to Calculate Expected Cell Frequencies 252
    Using Functions for a Chi-Square Test of Independence 254
    Interpreting the output..254

Chapter 13: Random Samples

13.1. Random Selection Using the Random Number Generation Tool 257
    Sample Research Problem 257
13.2. Random Selection Using the Sampling Tool 260
    Sample Research Problem 260
13.3. Random Selection Using the RANDBETWEEN Function 262
    Sample Research Problem 262

Chapter 1

Getting Started

▶ Section 1.1 | Introduction and Overview

This manual describes how to use Excel to perform common statistical procedures. Detailed, step-by-step examples are provided to illustrate how to use formulas, functions, the Pivot Table, the Pivot Chart, and Data Analysis Tools for statistical applications. This manual is designed to be used along with a statistics textbook, and, therefore, provides only brief explanations of the analyses themselves.

This manual contains two major parts: 1) a general introduction to Excel that I hope will provide enough information for you to be able to use the Excel worksheet and 2) chapters that describe how to use Excel to carry out statistical analysis procedures.

The first three chapters introduce worksheets, and describe how to enter data, use simple formulas to manipulate data, save, retrieve, and print. Chapter 4 shows how to organize data using frequency distributions and frequency graphs. Chapter 5 presents descriptive statistics and an explanation of how to weight responses to adjust for survey nonresponse. Chapter 6 discusses probability distributions and the ways that Excel can provide information about them. Chapter 7 presents techniques for testing hypotheses about one-sample means, followed by Chapter 8, which explains how to use Excel to test hypotheses about the difference between two means. Analysis of variance for simple cases is covered in Chapter 9. The Pearson correlation coefficient and the Spearman rank correlation coefficient are presented in Chapter 10. Chapter 11 includes bivariate regression, multiple regression, curvilinear regression, and dummy coding of qualitative variables. Cross tabulations and the chi-square test of independence are discussed in Chapter 12. Finally, Chapter 13 presents random sampling procedures.

Versions of Excel

The Microsoft Office 2010 version of Excel was used when preparing this manual and all of the screens pictured in this manual are from that version. Many of the new features incorporated into Excel 2007 have been retained in Excel 2010. If you are using a version of Excel that is earlier than 2007, you will not find this manual very helpful. For example, the menu and toolbars from versions earlier than Excel 2007 have been removed, the Pivot Table feature has been changed substantially so that it is much easier to use, and charts can now be constructed with just a couple of mouse clicks.

The main differences between Excel 2010 and 2007 with respect to data analysis occur in the statistical functions. A number of very useful functions have been added. For example, in Excel 2007, the user could only select two functions associated with the t distribution, TDIST and TINV. In Excel 2010, the user can select from T.DIST, T.DIST.2T, T.DIST.RT, T.INV, and T.INV.2T, depending on whether a one- or two-tailed test is being carried out and, for one-tailed tests, whether the critical t value is in the left or right tail of the distribution.

Versions of Windows

The copies of screens shown in this book were taken from a PC using Windows XP. The screens will appear slightly different if you are using a different Windows version or a Mac. After you are operating comfortably within Excel, these differences should be minor. There will, however, be slight differences between Macs and PCs in the keys used for commands.

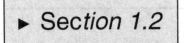 **What May Be Skipped**

If you have used spreadsheets before, you can probably omit much of the first three chapters. Other programs, such as Lotus 1-2-3 and Quattro, use somewhat different terminology to describe the operations covered in Chapters 2 and 3. The concepts, however, are essentially the same. For those of you who are looking for a specific procedure to use for a statistical test, each of the chapters can be used independently.

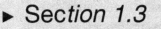 **Excel Worksheet Basics**

Using the Mouse

Mice come in several forms. The majority are provided with new computers and roll on a desktop or pad. When rolled, a small ball on the bottom causes the pointer on the screen (called the screen pointer) to move in a corresponding way. Another version (called a trackball) places a larger ball in a framework that allows you to roll the ball with your fingers. Other forms have small screens that you move your finger across as you move the mouse. The pressure of your finger moving across the screen causes a screen pointer to move in synchrony with your movements.

All devices have at least one button (and most likely two or more) that you can click or hold down, sometimes while also moving the ball. Five basic actions operate the mouse.

- **Point**. You point to objects on the screen by sliding the mouse on the deskpad or by rolling the trackball. As the screen pointer tracks the movements made on your desk, its shape changes. Most often the shape will be the outline of an arrow or the outline of a plus sign. The shape usually changes when a task is completed.

- **Click**. "Click" means to press and release the left mouse button (called a left-click). If you are pointing at an executable command, this action causes it to take place. If you point to any cell on the worksheet and click, that cell becomes the active cell and is ready to receive data.

- **Double-click**. "Double-click" means to press and release the left mouse button twice rapidly. If you fail to press rapidly enough, it is interpreted as one click. The double-click often replaces the two-step sequence of selecting a command and then clicking on OK to execute that command.

- **Right-click**. "Right-click" means to press down on the right button of the mouse. A right-click is often used to display special shortcut menus.

- **Drag**. Objects on the screen are moved by dragging. To drag, place the mouse pointer on the item you want to move, click and hold down the left mouse button—do NOT release it. While you hold the mouse button down, slide the mouse to move the screen pointer and the item to the location you want. Then release the mouse button.

Starting Excel

To start the program using Windows XP, click on **Start** in the lower left of the screen. Move the mouse to **Programs** and then continue moving through menus until you find the **Microsoft Excel** icon to click and begin.

Exiting Excel

There are two ways that you can exit Excel.

1. Click the upper **X** (the **close button**) that you find in the upper right corner of the screen. If two sets of boxes are showing, the lower set applies to the worksheet that is showing while the upper one is for the application or program itself (i.e., Excel). If you have edited (changed) any of the information in the workbook, you will be prompted to save the information before closing the program.

2. Click the **File** tab in the upper left corner. Select **Exit** in the bottom of the list on the left. You will be prompted to save the file if you edited any information in the workbook.

Layout of Worksheets

Across the top of the Excel 2010 worksheet, you see a row of several tabs: Home, Insert, Page Layout, Formulas, Data, Review, and View. Each tab leads to a ribbon. Each ribbon contains groups of related commands. The groups in the **Home ribbon**, displayed below, are Clipboard, Font, Alignment, Number, Styles, Cells, and Editing.

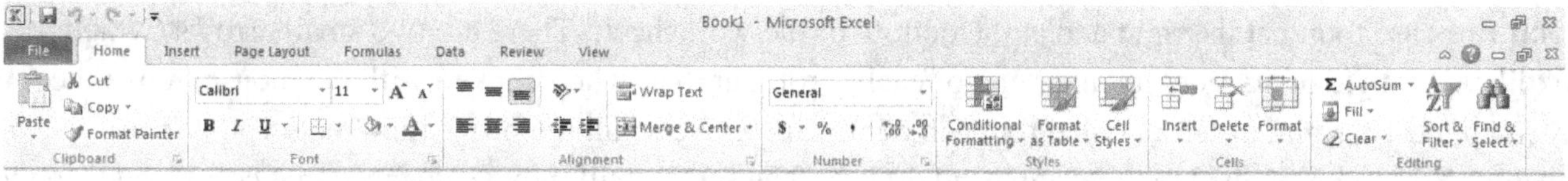

The **Insert ribbon**, shown below, includes groups of commands for Tables, Illustrations, Charts, Sparklines, Filter, Links, Text, and Symbols.

The **Page Layout ribbon** includes groups of commands for Themes, Page Setup, Scale to Fit, Sheet Options, and Arrange.

The **Formulas ribbon** includes groups of commands for Function Library, Defined Names, Formula Auditing, and Calculation. **Insert Function** is a command in the Function Library that you will be using a great deal.

The **Data ribbon** includes groups of commands for Get External Data, Connections, Sort & Filter, Data Tools, Outline, and Analysis. Many statistical analysis procedures are found in **Data Analysis** which is located in the Analysis group at the right end of the Data ribbon.

Worksheet Area. The worksheet consists of cells with columns labeled as letters and rows as numbers. Each cell is identified by the combination of its column letter and row number. Cell A1 is in column A, row 1. The dark border around cell A1 indicates that it is active and is ready to receive data.

Scroll Bars are found at the right and at the bottom of the worksheet. There are two small arrows (which look like a triangle on its side) at either end of each scroll bar that, when clicked with the mouse, move the screen up, down, right, or left one line for each click. Some users call the box within the scroll bar an elevator. You can grab the box (elevator) and drag it. The screen will move a distance that corresponds to the amount you move the box.

Worksheet Tabs. These are at the lower left of the screen and are labeled as Sheet 1, Sheet 2, and Sheet 3. You can insert additional worksheets, delete unwanted worksheets, and rename worksheets.

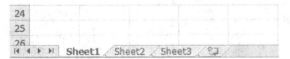

▶ Section 1.4 | Dialog Boxes

Dialog boxes usually require that you choose from a number of alternatives or enter your choices. Many of the statistical analysis procedures that are presented in this manual are associated with commands that are followed by dialog boxes. For example, if you click the **Formulas** tab near the top of the screen and select **Insert Function**, a dialog box like the one shown below will appear. You will usually select a function category and a function name. You make your selections by clicking on them.

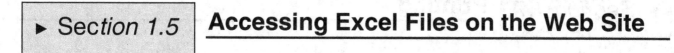

When you click the OK button at the bottom of the Insert Function dialog box, another dialog box will be displayed that asks you to provide information regarding location of the data in the Excel worksheet.

▶ Section 1.5 | Accessing Excel Files on the Web Site

Many of the data files that are used as examples in this manual are available on a Web site. The instructions at the beginning of an example will tell you the name of the file. The URL for the Web site is:

https://netfiles.umn.edu/users/dretz001/www/statistics/

After you open a file that you find on this Web site, you will want to save it on your computer.

▶ Section 1.6   Saving Information

Naming Workbooks

If you have opened a new workbook and entered data that you want to save, the next step is to click the **File tab** in the top left corner of the screen. The menu contains the commands **Save** and **Save As**.

If the sheet is new and has not yet been given a file name, you will be prompted to provide a name. The default name of **Book1** appears in the **File Name** window of the **Save As** dialog box. Unless you indicate otherwise, the file will be given the name Book1 and will be saved in the location last used, be it the hard drive, network server, or compact disk. A common practice is to save in a folder called **My Documents**.

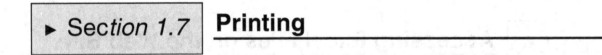

File names can have up to 255 characters including the extension. There are restrictions on the symbols that you can use in a file name, which will be apparent when you get an error message for using a / or some other forbidden symbol. If you give the file a name that indicates the worksheet contents, it will be easy to identify it when you want to return to it in the future. Some examples of possible file names are shown below.

June 2011 opinion survey

Project Expenses

MINNESOTA DATA

2007 Japan & China Info

▶ Section 1.7   Printing

Page Setup

Click the **Page Layout** tab at the top of the screen. Then click the arrow to the right of Page Setup. The **Page Setup** dialog box will appear. You will see four tabs along the top to access the four setup options: **Page**, **Margins**, **Header/Footer**, and **Sheet**.

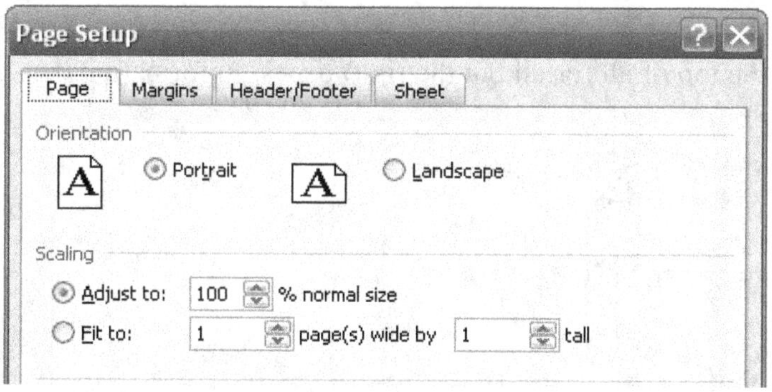

Page allows the user to change the **orientation** of the page from portrait (the way this book is printed) to landscape (sideways). With the **scaling** commands you can adjust the size of the image to be 10% of normal size or enlarge it up to 400%. The **Fit to** command instructs the program to automatically adjust the size to fit any of a number of pages as specified.

Click on the **Margins** tab to choose margin options.

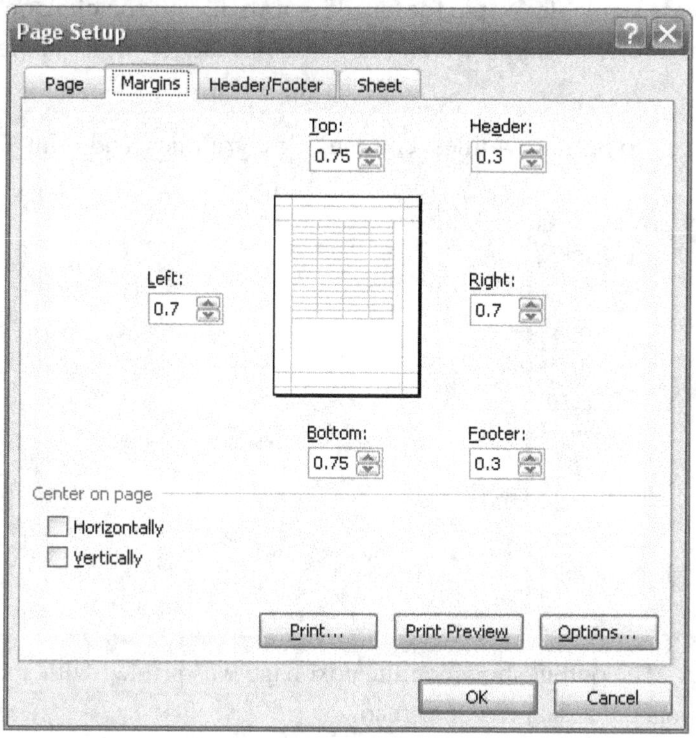

The preview window shows how the margins will look as you change them. Note that you can also center the output on the page horizontally, vertically, or both. After you have changed any of the settings, you can click on **Print Preview** to check the worksheet's appearance.

Headers contain the information printed across the top of all (or all but the first) pages. Information that is printed at the bottom of a page is called a **Footer**.

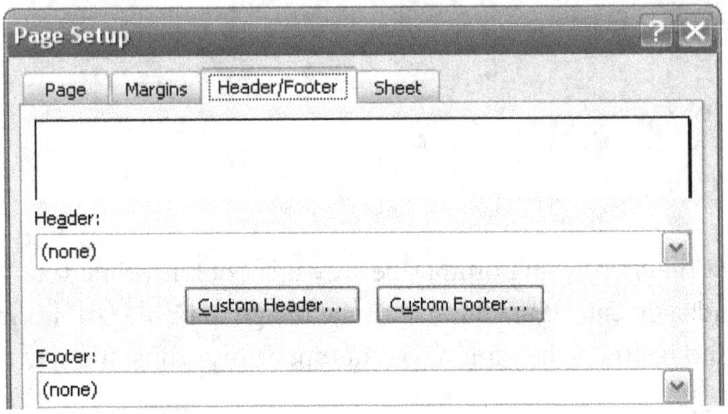

Page Setup

Page | Margins | Header/Footer | Sheet

Header:
(none)

Custom Header...   Custom Footer...

Footer:
(none)

Click on the arrow to the right of the Header window, and you will see several standardized options for the **header**. Similarly, several standardized options are available for the **footer**. If you wish, you can click on **Custom Header** or **Custom Footer** and type the information you would like for the header or footer in the spaces provided.

The **Sheet** dialog box allows you to choose the area to print, row titles, column titles, gridlines, and print quality.

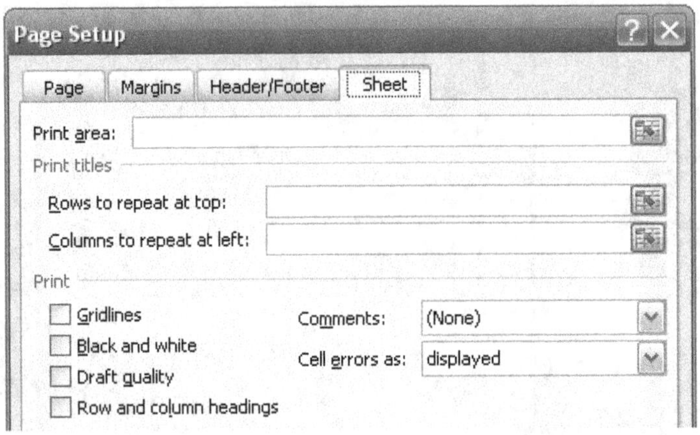

Page Setup

Page | Margins | Header/Footer | Sheet

Print area:

Print titles
Rows to repeat at top:
Columns to repeat at left:

Print
☐ Gridlines   Comments: (None)
☐ Black and white   Cell errors as: displayed
☐ Draft quality
☐ Row and column headings

The **Print** option frequently selected is **gridlines**. The output shown on the next page was printed with and without gridlines. As you can see, the gridlines make the data easier to read.

Treatment	Rating	X1	X2
Clean	6	1	0
Clean	5	1	0
Clean	7	1	0
Clean	6	1	0
Clean	3	1	0
Clean	7	1	0
Criminal	1	0	1
Criminal	2	0	1
Criminal	1	0	1
Criminal	3	0	1
Criminal	2	0	1
Control	4	0	0
Control	3	0	0
Control	5	0	0
Control	6	0	0
Control	4	0	0

Treatment	Rating	X1	X2
Clean	6	1	0
Clean	5	1	0
Clean	7	1	0
Clean	6	1	0
Clean	3	1	0
Clean	7	1	0
Criminal	1	0	1
Criminal	2	0	1
Criminal	1	0	1
Criminal	3	0	1
Criminal	2	0	1
Control	4	0	0
Control	3	0	0
Control	5	0	0
Control	6	0	0
Control	4	0	0

► Section 1.8 Loading Excel's Analysis Toolpak

The Analysis Toolpak is an Excel Add-In that may not necessarily be loaded on your computer. If **Data Analysis** does not appear in the Analysis group of the Data ribbon as shown below, then you will need to load it.

Data Analysis
Analysis

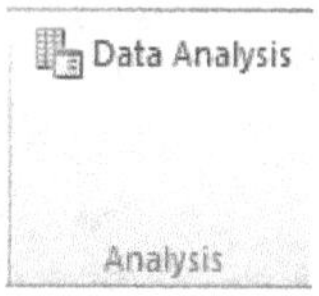

First click the **File** tab and select **Options**.

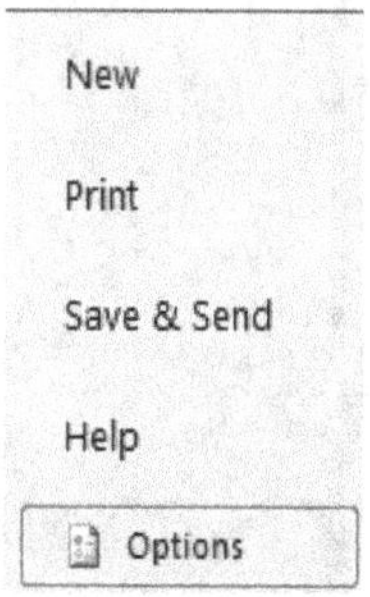

New
Print
Save & Send
Help
Options

Select **Add-Ins** from the list on the left.

General
Formulas
Proofing
Save
Language
Advanced
Customize Ribbon
Quick Access Toolbar
Add-Ins

Analysis ToolPak and Analysis ToolPak-VBA should both be in the list of Active Application Add-Ins. If you need to add one or both of them, first click on the name to select it. Then click **Go** at the bottom of the dialog box.

View and manage Microsoft Office Add-ins.

Add-ins

Name ▲	Location	Type
Active Application Add-ins		
Acrobat PDFMaker Office COM Addin	C:\....0\PDFMaker\Office\PDFMOfficeAddin.dll	COM Add-in
Analysis ToolPak	C:\...ce\Office14\Library\Analysis\ANALYS32.XLL	Excel Add-in
Inactive Application Add-ins		
Analysis ToolPak - VBA	C:\...Office14\Library\Analysis\ATPVBAEN.XLAM	Excel Add-in
Custom XML Data	C:\...es\Microsoft Office\Office14\OFFRHD.DLL	Document Inspector
Date (XML)	C:\...es\Microsoft Shared\Smart Tag\MOFL.DLL	Action
Euro Currency Tools	C:\... Office\Office14\Library\EUROTOOL.XLAM	Excel Add-in
Financial Symbol (XML)	C:\...es\Microsoft Shared\Smart Tag\MOFL.DLL	Action
Headers and Footers	C:\...es\Microsoft Office\Office14\OFFRHD.DLL	Document Inspector
Hidden Rows and Columns	C:\...es\Microsoft Office\Office14\OFFRHD.DLL	Document Inspector
Hidden Worksheets	C:\...es\Microsoft Office\Office14\OFFRHD.DLL	Document Inspector
Invisible Content	C:\...es\Microsoft Office\Office14\OFFRHD.DLL	Document Inspector
Microsoft Actions Pane 3		XML Expansion Pack
Smart Document for Customer Credit Memos		XML Expansion Pack
Smart Document for Customer Statements		XML Expansion Pack
Smart Document for Purchase Orders		XML Expansion Pack
Smart Document for Sales Invoices		XML Expansion Pack
Smart Document for Sales Orders		XML Expansion Pack
Smart Document for Sales Quotes		XML Expansion Pack
Solver Add-in	C:\...ce\Office14\Library\SOLVER\SOLVER.XLAM	Excel Add-in

Add-in:	Analysis ToolPak - VBA
Publisher:	
Compatibility:	No compatibility information available
Location:	C:\Program Files\Microsoft Office\Office14\Library\Analysis\ATPVBAEN.XLAM
Description:	VBA functions for Analysis ToolPak

Manage: Excel Add-ins   ▼   Go...

OK   Cancel

In the Add-Ins dialog box, place a check mark in the box next to the add-in to make it active. Click **OK**.

Chapter 2

Entering, Editing, and Recoding Information

Section 2.1 | Opening Documents

Opening a Brand-New Worksheet

If you started Excel by clicking on the icon, the screen opened with a new, blank worksheet. The title at the top indicates it is called **Book 1**. Along the bottom are **sheet tabs** labeled **Sheet 1**, **Sheet 2**, and **Sheet 3**. The sheets are stored together in one unit, called a book. You might, for example, choose to keep all data analysis for one research project together in one unit, now called **Book 1**, but renamed by you as **Opinion Survey**. You could create other books as well that contain the data and analyses for other research projects, records of project expenses, course assignments, and so on.

If you click on the X in the upper right of the Excel window, indicating that you want to quit, the program will ask you if you want to save the file you have created. If you have already saved it, and therefore have named it, Excel will save it using the same name unless you use the **Save As** command, which is used to change the name or location for saving. The Save As option is located in the menu that appears when you click the File tab located in the upper left corner of the screen.

Opening a File You Have Already Created

If you want to work on a file that you previously created, you will first click the **File** tab located in the upper left corner. Select **Open** from the menu.

Then locate and select the file.

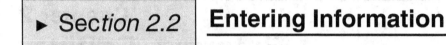

▶ Section 2.2   Entering Information

Addresses

The Excel 2010 grid has 16,384 columns. Columns are identified by letters of the alphabet that are shown across the top of the worksheet. Two letter combinations are used (e.g., AA, AB) are used after Z is reached, and then three letter combinations are used. The final column is labeled XFD. Rows continue numerically until the last row (1,048,576) is reached. Each cell is identified by a combination of a letter and a number. For example, cell A1 is located in column A, row 1, and cell D15 is located in column D, row 15.

Activating a Cell or Range of Cells

When the worksheet is initially opened, cell A1 is automatically the **active cell**. It has a dark outline around it, which indicates that whatever you type will be entered into that cell. The address A1 is displayed in the **Name Box,** which is to the left of the **Formula Bar**.

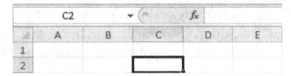

Move to cell C2, click, and note that C2 is displayed in the Name Box.

We often need to refer to more than one cell at a time. A group of cells is called a **range**. Click on cell B4, hold the mouse button down, and drag to B8. Release the button. The worksheet is now **marked** (highlighted) in the range of cells as shown below. Note that the top cell of the range is not darkened. To indicate the addresses of a range of cells, we separate the address of the upper left and the lower right cell with a colon. Here we have a range for a single column identified as **B4:B8**, although the Name Box displays only the address of the top cell (B4).

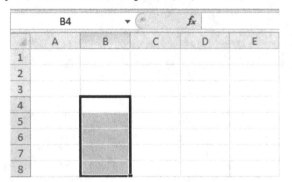

To activate cells in many rows and columns (i.e., a range), place the cursor in the upper left cell and drag to the lower right. Now all of the cells in that range will appear highlighted. You can click and drag in the opposite direction if you wish; that is, you can drag from the bottom to the top. This is a good way to proceed if you tend to overshoot your target and move past where you want to end, as many of us do. Another way to activate a range of cells is to click in the upper left cell, press and hold down the **SHIFT** key, and then move to the lower right cell using the scroll bars or arrow keys.

Types of Information

There are three types of information that may be entered into a cell:

1. **Text**. This term refers to alphabet characters (e.g., Exam) or a combination of alphabet, numeric, and symbol characters (e.g., Exam #2).

2. **Numeric**. Any cell entry comprised completely of numbers falls into the numeric category.

3. **Formulas**. These create new information generated from operations performed on numbers that are entered directly in the formula or that are entered in cells of the worksheet.

Filling Adjacent Cells

Filling can mean one of two things. First, filling can mean that you take the contents of a given cell or range of cells and make copies of that material in adjacent cells. Second, filling can also mean that you continue a series or sequence into adjacent cells.

Filling the Same Content into Adjacent Cells

If you have three different treatment conditions and you wish to repeat them in the first column of a worksheet, you would proceed as follows.

1. In a new worksheet, enter **Drug A** in cell A1, **Drug B** in cell A2, and **Placebo** in cell A3.

	A
1	Drug A
2	Drug B
3	Placebo

2. Click in cell A1 and drag to cell A3, to activate that range.

	A
1	Drug A
2	Drug B
3	Placebo

3. Use the **fill handle** to copy the contents of A1:A3 to cells A4 through A12. To do this, move the mouse pointer to the lower right corner of cell A3. It will turn into a black plus sign. Press the left mouse key and hold it down while you drag to cell A12. Release the mouse key and the names of the three conditions will be repeated in column A as shown at the top of the next page.

	A
1	Drug A
2	Drug B
3	Placebo
4	Drug A
5	Drug B
6	Placebo
7	Drug A
8	Drug B
9	Placebo
10	Drug A
11	Drug B
12	Placebo

Filling a Series into Adjacent Cells

The subject numbers assigned to research participants are often in numerical order, such that the first subject is 1, the second is 2, and so on.

1. Enter **1** in cell A1 and **2** in cell A2.

	A
1	1
2	2

2. Mark cells A1 and A2 by dragging over them.

	A
1	1
2	2

3. Use the fill handle to continue the series in cells A3 through A10.

	A
1	1
2	2
3	3
4	4
5	5
6	6
7	7
8	8
9	9
10	10

Series

The previous activities show how easily you can repeat cell content in adjacent cells or fill adjacent cells with a series of numbers. The **Series** command can also be used generate a series of numbers. Follow these steps to practice using this command.

1. Enter **5**, **7**, and **9** in cells A1, A2, and A3, respectively.

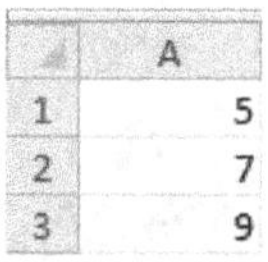

◢	A
1	5
2	7
3	9

2. Activate the range **A1:A10**.

3. Click the down arrow to the right of Fill in the Editing group of the Home ribbon. Select **Series**.

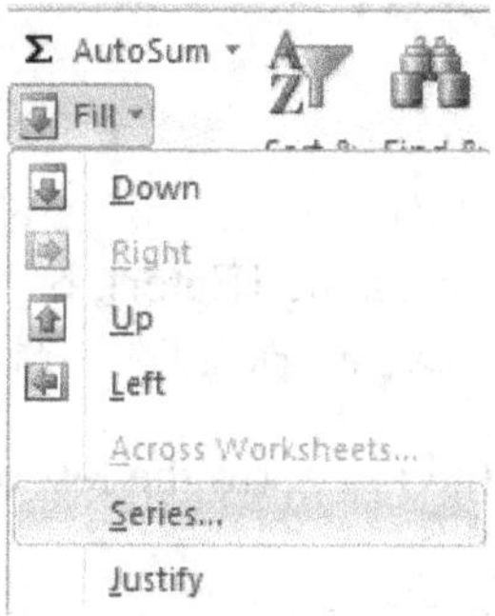

Σ AutoSum ▾

⬇ Fill ▾

⬇ **D**own

⬈ **R**ight

⬆ **U**p

⬅ **L**eft

Across Worksheets...

Series...

Justify

4. Excel has pre-selected **Columns** and **Linear**. In addition, Excel has correctly inferred that the step value is 2. You don't need to put in a stop value because you have marked the range of cells that are to be filled. Click **OK**.

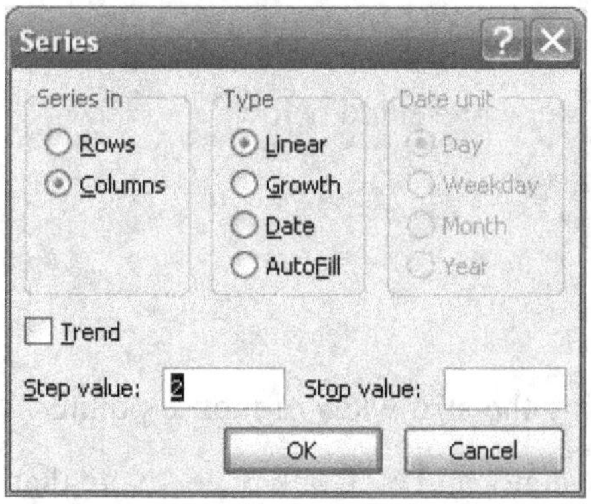

Series   [?] [✕]

Series in   Type   Date unit

○ **R**ows   ⦿ **L**inear   ⦿ **D**ay

⦿ **C**olumns   ○ **G**rowth   ○ **W**eekday

                   ○ **D**ate   ○ **M**onth

                   ○ **A**utoFill   ○ **Y**ear

☐ **T**rend

Step value: 2   Sto**p** value: [   ]

                   [OK]   [Cancel]

The marked range in the worksheet now contains the series of odd numbers beginning with 5 and ending with 23

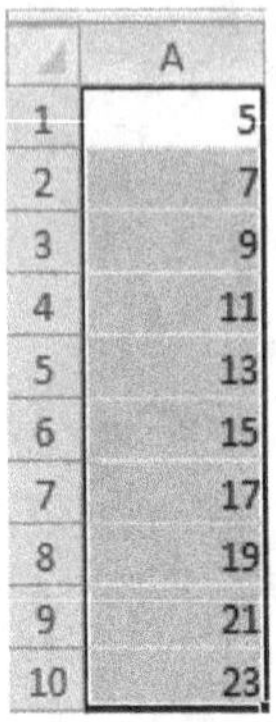

	A
1	5
2	7
3	9
4	11
5	13
6	15
7	17
8	19
9	21
10	23

▶ Section 2.3 | Editing Information

Changing Information

To change information in a cell, you have to consider which situation exists.

- If you have not yet "accepted" the information by clicking the green check mark (or pressing [Enter], or using arrow keys, or …), then you can simply use the backspace or delete key to remove entries.

- If you are typing information into a cell and decide you want to start over, click on the red X (to the left of the Editing Bar) and everything will be deleted.

- If you want to delete everything in the active cell or range of cells, press the [Delete] key.

- If you have already entered and accepted data in a cell, but now want to edit it but not erase all of it, first activate the cell. Then, in the Editing Bar, move the mouse pointer to where you want to make the changes, and click to insert the I-beam.

As a practice exercise for changing information, try going through the steps presented below.

1. In cell A1 type **12346**. Press [**Enter**], which moves you to cell A2.

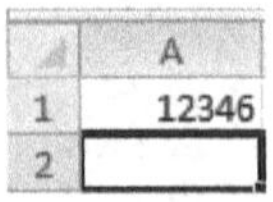

	A
1	12346
2	

2. Assume you really wanted to enter 123456. Return to **A1** by using the arrow key or mouse pointer.

3. In the Editing Bar, move the mouse pointer so the I-beam is between 4 and 6. Click once. Note that three editing keys are now shown to the left of 12346.

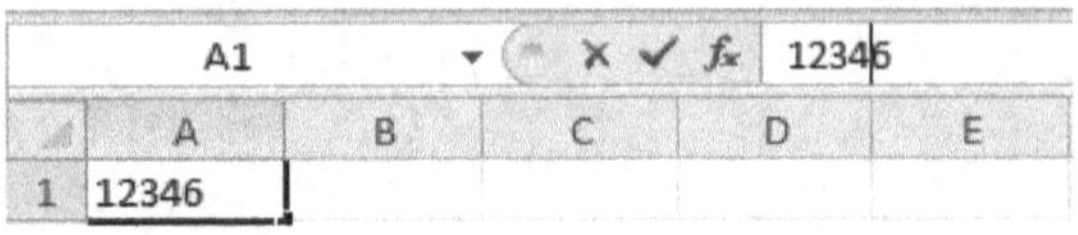

A1	▾	X ✓ fx	12346		
	A	B	C	D	E
1	12346				

4. Type **5**, which will be inserted between 4 and 6.

5. Press [**Enter**].

Moving and Copying Information

A basic principle is that you first indicate which material will have something done to it by marking it, and then you execute the command that does something to the highlighted material. You will see this principle operate in several other places in Excel. When you want to move information, you can do it so that it is removed from one location and placed in another. This is a **cut**. If you want to make a copy of the material so that it is in the original location as well as in another location, you make a **copy**.

1. To copy the entry in cell A1 (123456) to cell B1, first mark cell A1 by clicking on it.

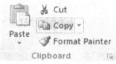

2. Move the mouse pointer to the Clipboard group of the Home ribbon and click on **Copy**.

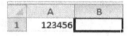

3. Click the cell where you want the copy placed, cell **B1**.

4. Move the mouse pointer to the Clipboard group of the Home ribbon and click on **Paste**.

5. The copy command operates like a rubber stamp; a copy is stored on the Clipboard and you can continue to place copies anywhere you wish. Activate another cell, say, C3, and paste another copy there.

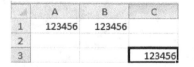

Cutting is done in a similar fashion, except that the cell is empty after you cut the data from it.

1. Click on a cell with content. For this example, click on cell **A1**.

2. Click on **Cut** in the Clipboard group of the Home ribbon.

3. Activate the cell where you would like the information to be placed. For this example, click in cell **A3**.

	A
1	123456
2	
3	

4. Move the mouse pointer to the Clipboard group of the Home ribbon and click on **Paste**.

	A
1	
2	
3	123456

If you cut or copy a group of cells (i.e., a range), the principle is the same. Instead of activating one cell, you click and drag over the cells so that a range of cells is marked. Pasting a range is the same, except that you specify only the upper left cell of the target location.

Moving material between worksheets is accomplished the same way: Mark, indicate Cut or Copy, move to the new sheet and the desired location for the upper left cell, and then paste.

Moving and Copying an Entire Worksheet

If you want to move or copy an entire worksheet, follow these steps.

1. Activate a cell in the worksheet that you want to copy. Then click on **Format** in the Cells group of the Home ribbon.

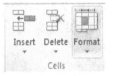

2. Select **Move or Copy Sheet** from the menu that appears.

Cell Size
- Row Height...
- AutoFit Row Height
- Column Width...
- AutoFit Column Width
- Default Width...

Visibility
- Hide & Unhide   ▸

Organize Sheets
- Rename Sheet
- Move or Copy Sheet...

3. To copy the worksheet, click in the box next to **Create a copy**. To move the worksheet, select from the options that are provided. In this example, a copy will be created and it will be moved to the end. Click **OK**.

Move or Copy   ? ✕

Move selected sheets
To book:

Book1   ▾

Before sheet:

Sheet1
Sheet2
Sheet3
(move to end)

☑ Create a copy

OK   Cancel

Dragging and Dropping

If you want to move or copy a range of cells on the same sheet, a shortcut is called **drag and drop**. To move a range, mark the cells and position the cursor near an edge, where it will become an outline arrow. Hold down the left key of the mouse and drag to the new location. Release the mouse key. To copy, hold down the **CTRL** key at the same time. If you attempt to drag and drop material into a spot that currently has data, you will get a message that asks if you want to do this, because the old material will be removed. If you have mistakenly done this, you can fall back on the **Undo** icon that you will find in the top left corner of the screen.

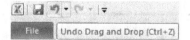

File   Undo Drag and Drop (Ctrl+Z)

Inserting or Deleting Rows and Columns

Inserting or deleting rows and columns is relatively easy. If you have entered data in a row that includes columns A, B, and C and want to place a new column between A and B, do the following. Click on the letter **B** to mark that column. Click **Insert** in the Cells group of the Home ribbon and select **Insert Sheet Columns**.

A column will be inserted to the left of the marked column. If you want to insert two columns between A and B, simply mark columns B and C at the same time and follow the procedure described above. To delete these two empty columns, mark them and click **Delete** in the Cells group of the Home ribbon and select **Delete Sheet Columns**.

The procedure for inserting or deleting rows is exactly the same, except that you use row numbers instead of column letters.

Changing the Column Width

Column width can be changed in a couple different ways. Only one way will be described here. Output from the Descriptive Statistics Analysis Tool will be used as an example. As you can see in the output displayed below, many of the labels in column A can only be partially viewed because the column width is too narrow.

	A	B
1	TV	
2		
3	Mean	2.636364
4	Standard I	0.777791
5	Median	2
6	Mode	3
7	Standard I	2.579641
8	Sample Vi	6.654545

Position the mouse pointer directly on the vertical line between A and B in the letter row at the top of the columns [A | B] so that it turns into a black plus sign. Click and drag to the right until you can read all the output labels. Release the mouse key. After adjusting the column width, your output should appear similar to the output shown below. If you have made the column too wide, you can decrease the width by clicking on the vertical line and then dragging to the left.

	A	B
1	TV	
2		
3	Mean	2.636364
4	Standard Error	0.777791
5	Median	2
6	Mode	3
7	Standard Deviation	2.579641
8	Sample Variance	6.654545

► Section 2.4   Formatting Numbers

Formatting options for numerical data are available in the Number group of the Home ribbon. Options for displaying decimals and commas are available as well as many special format categories, such as currency, accounting, date, time, and percentage. In this section, I will explain decimals, currency, and percentage.

The same worksheet will be used for all the examples. Start with a new worksheet. Enter the numbers shown below in cells A1 through A6, and then copy them into columns B, C, and D.

	A	B	C	D
1	1.234567	1.234567	1.234567	1.234567
2	2.34567	2.34567	2.34567	2.34567
3	3.4567	3.4567	3.4567	3.4567
4	4.567	4.567	4.567	4.567
5	5.67	5.67	5.67	5.67
6	6.7	6.7	6.7	6.7

Decimal Points

1. Click on the **B** at the top of the second column to mark it.

	A	B	C	D
1	1.234567	1.234567	1.234567	1.234567
2	2.34567	2.34567	2.34567	2.34567
3	3.4567	3.4567	3.4567	3.4567
4	4.567	4.567	4.567	4.567
5	5.67	5.67	5.67	5.67
6	6.7	6.7	6.7	6.7

2. Move the mouse pointer to the Number group of the Home ribbon. Click the down arrow next to General and select **Number**. The number of decimal places has been pre-selected so that two decimal points will be displayed.

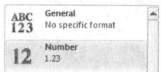

ABC 123	General No specific format
12	Number 1.23

3. If you would like a different number of decimal places, click the appropriate arrow in the Number group. One arrow will increase decimals and the other will decrease decimals. For this example, let's display 4 decimal places.

	A	B	C	D
1	1.234567	1.2346	1.234567	1.234567
2	2.34567	2.3457	2.34567	2.34567
3	3.4567	3.4567	3.4567	3.4567
4	4.567	4.5670	4.567	4.567
5	5.67	5.6700	5.67	5.67
6	6.7	6.7000	6.7	6.7

Currency

1. Click on **C** at the top of the third column to mark it.

	A	B	C	D
1	1.234567	1.2346	1.234567	1.234567
2	2.34567	2.3457	2.34567	2.34567
3	3.4567	3.4567	3.4567	3.4567
4	4.567	4.5670	4.567	4.567
5	5.67	5.6700	5.67	5.67
6	6.7	6.7000	6.7	6.7

2. Move the mouse pointer to the Number group of the Home ribbon. Click the down arrow next to General and select **Currency**.

ABC 123	General No specific format
12	Number 1.23
	Currency $1.23

The number of decimal places has been pre-selected so that two decimal points will be displayed.

	A	B	C	D
1	1.234567	1.2346	$1.23	1.234567
2	2.34567	2.3457	$2.35	2.34567
3	3.4567	3.4567	$3.46	3.4567
4	4.567	4.5670	$4.57	4.567
5	5.67	5.6700	$5.67	5.67
6	6.7	6.7000	$6.70	6.7

Percentage

1. Click on **D** at the top of the fourth column to mark it.

	A	B	C	D
1	1.234567	1.2346	$1.23	1.234567
2	2.34567	2.3457	$2.35	2.34567
3	3.4567	3.4567	$3.46	3.4567
4	4.567	4.5670	$4.57	4.567
5	5.67	5.6700	$5.67	5.67
6	6.7	6.7000	$6.70	6.7

2. Move the mouse pointer to the Number group of the Home ribbon. Click the down arrow next to General and select **Percentage** as shown at the top of the next page. The number of decimal places has been pre-selected so that two decimal points will be displayed.

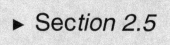

ABC 123	General No specific format
12	Number 1.23
	Currency $1.23
	Accounting $1.23
	Short Date 1/1/1900
	Long Date Sunday, January 01, 1900
	Time 5:37:47 AM
%	Percentage 123.46%

Note that the percentage format multiplies the cell value by 100. This format would be especially useful if your original cell entries were proportions, such as .123, .234, etc., and you wished to express them as percentages.

	A	B	C	D
1	1.234567	1.2346	$1.23	123.46%
2	2.34567	2.3457	$2.35	234.57%
3	3.4567	3.4567	$3.46	345.67%
4	4.567	4.5670	$4.57	456.70%
5	5.67	5.6700	$5.67	567.00%
6	6.7	6.7000	$6.70	670.00%

► Section 2.5   Recoding

Recoding refers to creating a new categorical variable for an existing variable that takes on numerical values. For example, you may have recorded age in years and now want to analyze age as a categorical variable with categories of young, middle, and older.

1. Enter the Age values shown at the top of the next page. These values represent number of years.

	A
1	Age
2	18
3	38
4	21
5	45
6	52
7	60
8	68
9	37
10	40
11	35
12	29
13	30

2. You want to recode Age into a categorical variable where less than 30 years will be *Young*, 30 to 59 years will be *Middle*, and more than 59 years will be *Older*. Create a new variable with a recognizable label such as Age Group. Enter the label **Age Group** in cell **B1**.

	A	B
1	Age	Age Group
2	18	

3. You will use the VLOOKUP function to give values to Age Group. The VLOOKUP function requires a reference table. Key in the reference table as shown below in the range E1:F3. Then click in cell **B2** to assign an Age Group to age of 18 years.

	A	B	C	D	E	F
1	Age	Age Group			0	Young
2	18				30	Middle
3	38				60	Older

4. Click the **Formulas** tab near the top of the screen and select **Insert Function**.

5. Select the **Lookup & Reference** category. Select the **VLOOKUP** function. Click **OK**.

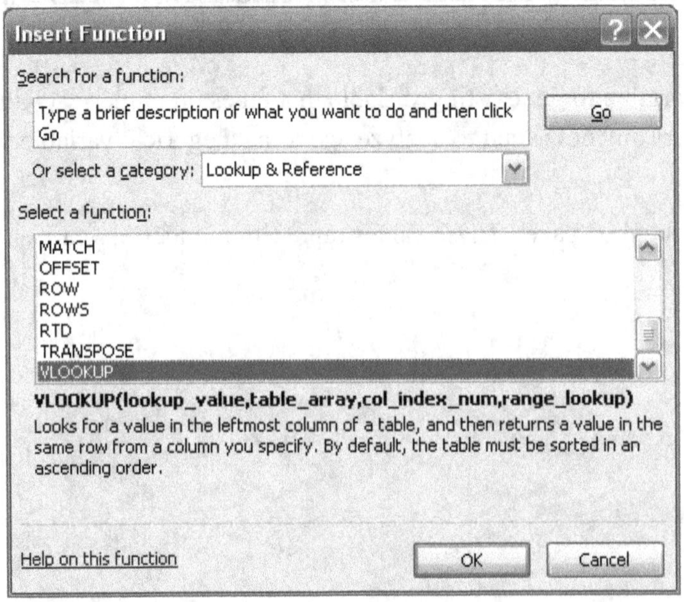

6. Complete the Function Arguments dialog box as shown below. The dollar signs in the table array window are necessary because the cell references must be absolute so that the correct values are returned when you copy the function. Click **OK**.

This VLOOKUP function says to take the value in cell A2 and compare it to the numbers in the first column of the reference table. Then assign an age group category from the second column of the reference table. If the number is equal to or greater than 0 and less than 30, then assign the "Young" category. If the number is equal to or greater than 30 and less than 60, then assign the "Middle" category. If the number is equal to or greater than 60, then assign the "Older" category.

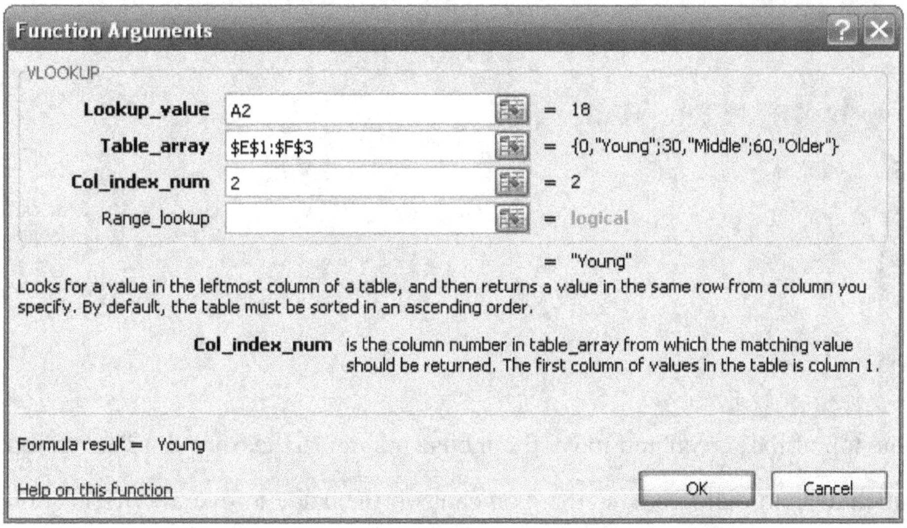

	A	B
1	Age	Age Group
2	18	Young

7. The VLOOKUP function correctly returns the Young category for the age of 18 years. Copy the function in B2 to cells B3:B13.

The completed worksheet is shown below.

	A	B	C	D	E	F
1	Age	Age Group			0	Young
2	18	Young			30	Middle
3	38	Middle			60	Older
4	21	Young				
5	45	Middle				
6	52	Middle				
7	60	Older				
8	68	Older				
9	37	Middle				
10	40	Middle				
11	35	Middle				
12	29	Young				
13	30	Middle				

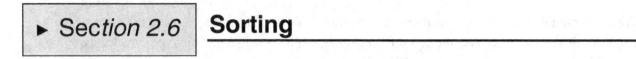

► Sec*tion 2.6* | **Sorting**

Sorting is a useful tool for becoming familiar with a data set that will be analyzed. You can sort data in ascending or descending order. You also can sort by two or more variables.

1. Enter the data as displayed in the worksheet below.

	A	B	C
1	Name	Weight	Eyes
2	Fred	200	Blue
3	Sue	147	Brown
4	John	185	Brown
5	Jane	98	Blue

2. Mark the range **A1 to C5**.

	A	B	C
1	Name	Weight	Eyes
2	Fred	200	Blue
3	Sue	147	Brown
4	John	185	Brown
5	Jane	98	Blue

3. Click the **Data** tab near the top of the screen and move the mouse pointer to the **Sort & Filter** group.

4. Click the Sort icon to launch the sort dialog box so that you can sort the data on several criteria at once.

5. Click the down arrow next to the Sort by window. Select one of the three variables as the first sorting variable. Let's use **Name** for the first sorting variable. You have the choice of sorting in A to Z order (ascending) or Z to A order (descending). Let's sort in **A to Z** order. Check to be sure that there is a check mark in the box next to **My data has headers** located in the top right of the dialog box. You don't want the labels *Name*, *Weight*, and *Eyes* to be included in the sort.

Sort			? ☒
⊞ Add Level ✕ Delete Level 🗐 Copy Level ▲ ▼ Options...		☑ My data has headers	
Column	Sort On	Order	
Sort by Name ⌄	Values ⌄	A to Z ⌄	
		OK Cancel	

6. You can sort by more than one variable. Click the **Add Level** button at the top. Select **Eyes** to be sorted in A to Z order.

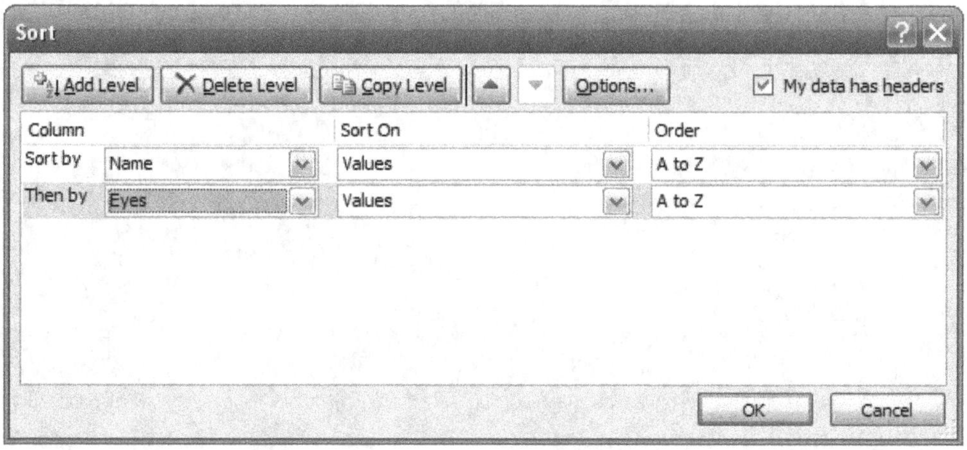

7. Click **OK**, and the data will be sorted as shown in the worksheet below.

	A	B	C
1	Name	Weight	Eyes
2	Fred	200	Blue
3	Jane	98	Blue
4	John	185	Brown
5	Sue	147	Brown

Chapter 3
Formulas

► Sec*tion 3.1*	**Operators**

Algebraic formulas utilize the four common mathematical operations of addition (+), subtraction (-), multiplication (x), and division (÷). When creating formulas, we will use these four operations and three others: negation, exponentiation, and percent. In Excel, these **operators** are symbolized as follows:

> Addition +
>
> Subtraction –
>
> Multiplication *
>
> Division /
>
> Negation, using the minus sign to indicate a negative number, as in –3
>
> Exponentiation ^
>
> Percent %

Multiplication is indicated by an asterisk (*) rather than an x. Division is indicated by a diagonal (/) instead of the division symbol. If we wish to raise a value to a power, say, X squared, we place a carat (^) between the X and the 2. In Excel, the formula for X squared appears as X^2. We can place a percent sign (%) after a value, as in 25%. For example, the formula **=15^2*25%** raises 15 to the second power and multiplies the result by .25 (the decimal form of 25%) to produce the result of 56.25.

Operations can be grouped together within parentheses. The parentheses determine the order in which the commands will be executed by Excel. Any operations enclosed in parentheses are executed first, from the innermost parentheses to the outermost.

Order of Operations

The order of operations in Excel is:

1. Negation
2. Percent
3. Exponentiation
4. Multiplication and division
5. Addition and subtraction

Excel first calculates expressions in parentheses and then uses those results to complete the calculations of the formula. Examples are given at the top of the next page.

=2+4*3   produces 14 because multiplication occurs before addition.

=(2+4)*3   produces 18 because operations within parentheses are executed first.

=2*(4+3)   produces 14 because operations within parentheses are executed first.

In a more complex formula,

=2+(4*(3+5)^2)/2   produces 130.

The more complex formula is carried out as follows: First, the terms in the innermost parentheses are executed, producing 8. Next, the 8 is raised to the second power, producing 64, since this is within the second set of parentheses and exponentiation has precedence over the other operations. Then 64 is multiplied by 4, yielding 256. The 256 is then divided by 2, producing 128. Finally, 2 is added to 128, yielding 130. Try this on your computer by keying the formula into cell A1. Press [**Enter**] when the formula is complete.

Writing Equations

In the equations above, we always used numeric constants such as 2, 3, 4, or 5. With Excel, the formulas that you write will more likely contain an address for a cell that can hold almost any value. This is equivalent to the X and Y that act as unknowns in algebra. For example, to convert degrees Fahrenheit to degrees Celsius, subtract 32 from the Fahrenheit temperature, multiply by 5, and divide by 9. In Excel, this formula would begin with an equal sign and would appear as **=(F-32)*5/9**, where F is the Fahrenheit temperature that we wish to convert to Celsius. Let's set up a table to do this conversion.

1. Type the label **Fahrenheit** in cell **A1**. Type **Celsius** in **B1.**

2. Enter the two values of **65** and **64** in cells **A2** and **A3** as shown in the worksheet below.

	A	B
1	Fahrenheit	Celsius
2	65	
3	64	
4		

3. Let Excel continue the Fahrenheit series. Click on **65** and drag down to cell **A80** to mark those cells. Click the **Home** tab near the top of the screen and click **Fill** in the Editing group.

4. Select **Series**.

Down
Right
Up
Left
Across Worksheets...
Series...
Justify

5. Notice that Excel has correctly inferred that the step size is –1. We want to continue the series using that one-unit change. Click **OK** and the temperatures extend from 65 to –13 degrees Fahrenheit.

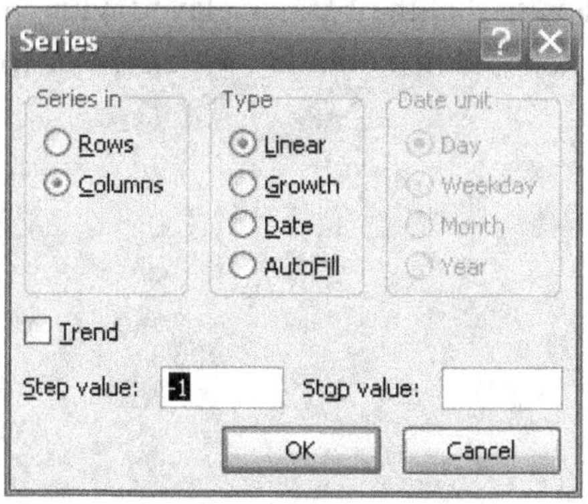

6. Let's calculate degrees Celsius using the conversion formula and place the results in column B. Activate cell **B2** by clicking in it.

7. We will use A2 in the formula. A2 is the cell address of 65 degrees Fahrenheit. Enter the formula **=(A2-32)*5/9**.

	A	B	C
1	Fahrenheit	Celsius	
2	65	=(A2-32)*5/9	
3	64		

8. Press [**Enter**]. You should now see 18.3333 in cell B2.

	A	B
1	Fahrenheit	Celsius
2	65	18.33333
3	64	

9. Next we will copy the formula in cell B2 down to cell B80. Activate cell **B2**. Then click on the small square in the lower right of cell B2 (the fill handle) and drag to cell B80. The screen pointer becomes a solid black plus sign when you use the fill handle. If these steps have been carried out correctly, your worksheet should look like the one shown below. The address of each A column cell that contained degrees Fahrenheit was changed in the formula as it was copied into the cells of column B. Click in cell **B3** and look in the Formula Bar to check the contents of the cell. You should see =(A3-32)*5/9.

B3			f_x	=(A3-32)*5/9	
	A	B	C	D	E
1	Fahrenheit	Celsius			
2	65	18.33333			
3	64	17.77778			
4	63	17.22222			
5	62	16.66667			
6	61	16.11111			
7	60	15.55556			

10. To make our output more attractive, let's format the Celsius values so that they are displayed with two decimal places. Click on letter **B** at the very top of column B to activate that column. Click the down arrow next to General in the Number group of the Home ribbon.

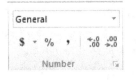

11. Select the **Number** category by clicking on it.

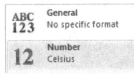

12. The first few rows of your worksheet should now appear like the ones shown below. The default display is two decimal places. You can easily increase or decrease the number of decimal places by clicking the appropriate arrow in the Number group of the Home ribbon. Click the arrow on the left to increase decimal places. Click the arrow on the right to decrease decimal places.

	A	B
1	Fahrenheit	Celsius
2	65	18.33
3	64	17.78
4	63	17.22
5	62	16.67

Relative References

When we entered the formula for determining Celsius degrees in cell B2 for our temperature conversion example, we entered it using the relative reference mode. We told the program to go to cell A2 and obtain the value located there to use in the formula. Actually, we instructed Excel to move one cell location to the left of the active cell and obtain the value that was stored there. So, when the formula in B2 was copied to B3, Excel moved one cell to the left to obtain the Fahrenheit value in A3, and so on. In other words, when the cell address is relative, Excel moves to a cell address relative to its starting position. Relative addresses are written without dollar signs. For example, B2 is a relative address.

Absolute References

Regardless of the starting point, absolute references do not change. For example, let's say that we are given the monthly sales of candy bars at three convenience stores: $300, $560 and $725. We want to know the average daily sales. If there are 30 days in each of these months, we would divide monthly sales by 30 to find the daily average. In this example, 30 is a constant, meaning that it does not change. Similarly, if the cell address of 30 were used instead of the actual value, the computing formula would need to express the cell address in such a way that would indicate that it does not change. In other words, the cell address needs to be absolute. Absolute cell addresses are written with dollar signs. For example, $B$2 is an absolute address. Work through this short example to see how to use an absolute address.

1. Enter the monthly sales of **300, 560,** and **725** in the first three cells of column A as shown below, and enter **30** in cell A6.

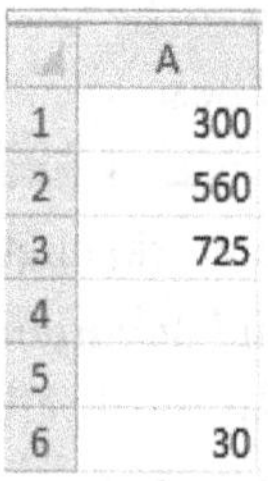

	A
1	300
2	560
3	725
4	
5	
6	30

2. Now let's compute average daily sales using a formula. Activate cell **B1**. Key in this formula to compute daily sales: **=A1/$A$6**.

The period at the end of the sentence is not part of the formula.

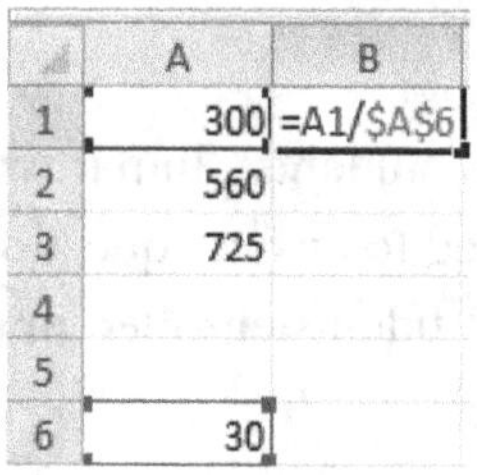

	A	B
1	300	=A1/$A$6
2	560	
3	725	
4		
5		
6	30	

3. Press [**Enter**]. You should now see 10 in cell B1.

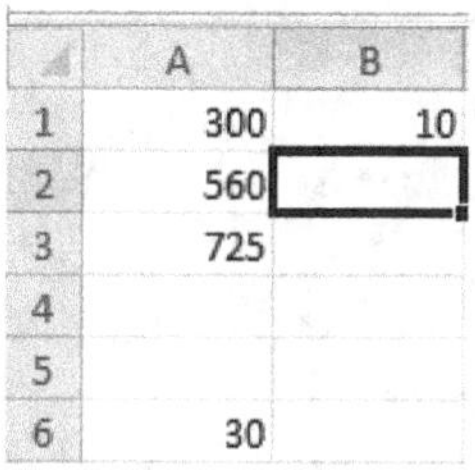

	A	B
1	300	10
2	560	
3	725	
4		
5		
6	30	

4. Let's copy the formula in B1 to cells B2 and B3. Activate cell **B1**. Then click on the small square in the lower right of cell B1 (the fill handle) and drag to cell B3. As you can see, the values in cells A2 and A3 were each divided by 30.

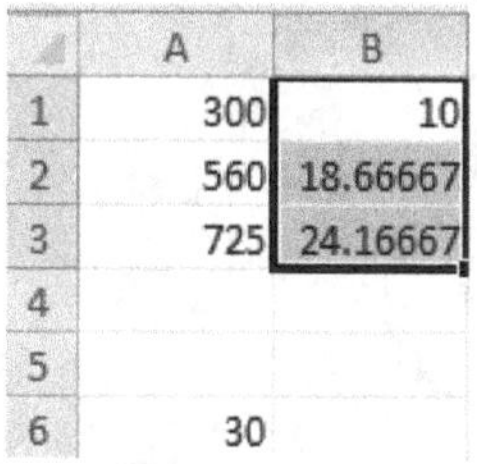

	A	B
1	300	10
2	560	18.66667
3	725	24.16667
4		
5		
6	30	

5. Click in cell B2 and look at the Formula Bar. You will see A2 and $A$6. Then click in cell B3 and look at the Formula Bar. You will see A3 and $A$6. The first cell reference is relative—it changes when the formula is copied down from one row to another. The second reference is absolute—it does not change.

Depending on the nature of the calculation you need to make, you can make both the column and row of a cell reference absolute (e.g., $A$1), only the column (e.g., $A1), or only the row (e.g., A$1).

▶ Section 3.2	**Using Formulas in Statistics**

For beginning statistics students, many analyses are carried out using a hand-held calculator. These calculations could also be performed using Excel. An important advantage associated with the use of Excel is that the student will know whether or not a number was entered correctly just by looking at the worksheet. Another important advantage is that the student will be able to carry out calculations quickly and accurately by copying a formula in one cell to other cells. In this section, I will present instructions for using formulas to calculate the mean, deviation scores, squared deviation scores, variance, standard deviation, and z-scores. The sections are presented in the same order in which the calculations would be carried out if you were doing them by hand. If you are using Excel to do the examples shown in this section of the manual, you will need to start with mean, then deviation scores, then squared deviation scores, and so on, ending with z-scores.

Sample Research Problem

A researcher was interested in describing the classroom behavior of gifted high school students. In particular, the researcher was interested in the number of times that the gifted students volunteered to answer questions or make comments in a science class. At the end of one class session, the five gifted students enrolled in the course received the participation scores shown below.

	A	B
1		Score
2		3
3		14
4		0
5		17
6		6

Mean

1. Enter the five participation scores in column B of an Excel worksheet as shown above. We'll use column A for labels.

2. We will begin by calculating the sum of the five scores. First, activate cell **A7** and key in **Sum**.

	A	B
1		Score
2		3
3		14
4		0
5		17
6		6
7	Sum	

3. Click in cell **B7**. Then click **AutoSum** near the top of your screen. AutoSum is found in the Editing group of the Home ribbon. AutoSum automatically enters the equal sign, the SUM function, and the range of numerical values immediately above the activated cell.

Instead of using the AutoSum button, you could type **=Sum**.

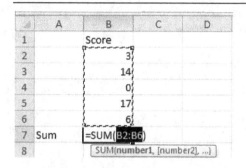

4. Check to make sure that the range is accurate. If it is not, make the necessary corrections. When the range is correct, press [**Enter**]. You should see 40 in cell B7.

	A	B
1		Score
2		3
3		14
4		0
5		17
6		6
7	Sum	40

5. The mean is computed by dividing the sum by the number of observations. Let's use Count to refer to the number of observations. Activate cell **A8** and key in **Count**.

6. Activate cell **B8**. We will now use Excel's COUNT function to provide the number of observations.

	A	B
1		Score
2		3
3		14
4		0
5		17
6		6
7	Sum	40
8	Count	

7. At the top of the screen, click the **Formulas** tab and select **Insert Function**.

8. In the Insert Function dialog box, select the **Statistical** category and the **COUNT** function. Click **OK**.

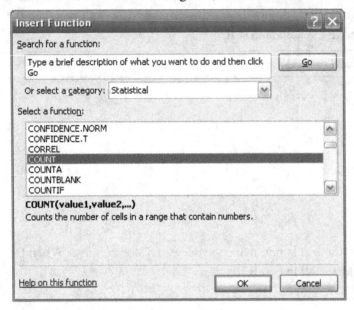

9. In the Function Arguments dialog box, you should see B2:B7. This needs to be changed to B2:B6, because you do not want to include the sum in the count.

10. Click in the **Value 1** window and change B7 to **B6**. Click **OK**.

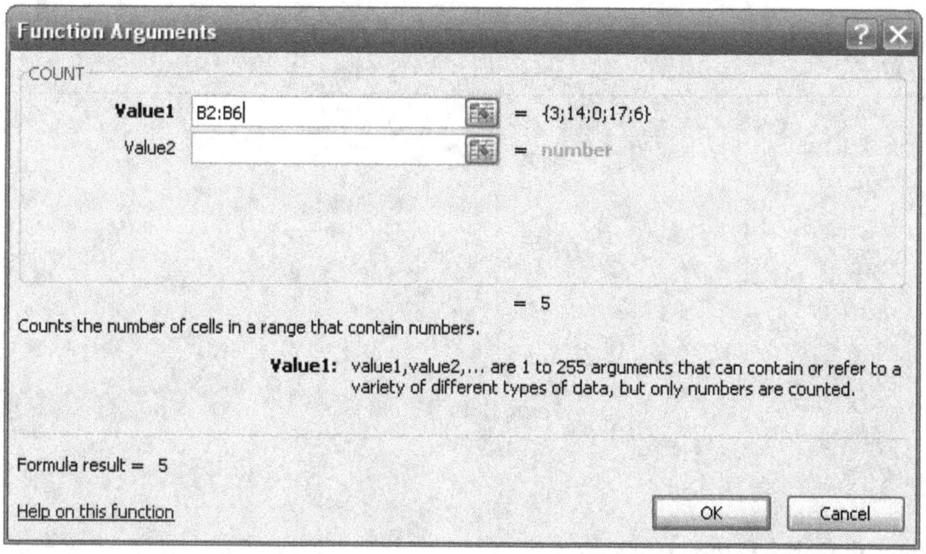

Function Arguments		? X

COUNT

Value1	B2:B6	= {3;14;0;17;6}
Value2		= number

= 5

Counts the number of cells in a range that contain numbers.

Value1: value1,value2,... are 1 to 255 arguments that can contain or refer to a variety of different types of data, but only numbers are counted.

Formula result = 5

Help on this function   OK   Cancel

11. You should now see 5 in cell B8. Activate cell **A9** and type **Mean**. Activate cell B9.

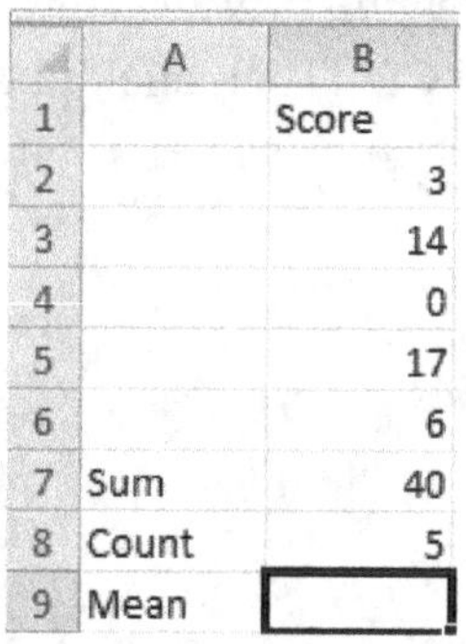

	A	B
1		Score
2		3
3		14
4		0
5		17
6		6
7	Sum	40
8	Count	5
9	Mean	

12. To compute the mean, we divide Sum by Count. Using cell addresses, this formula would be expressed as =B7/B8. The recommended procedure is to click in the cell addresses to make them part of the formula. Type = in cell B9 and then click in cell **B7**. Type **/** and then click in cell **B8**. In the Formula Bar, you should see =B7/B8.

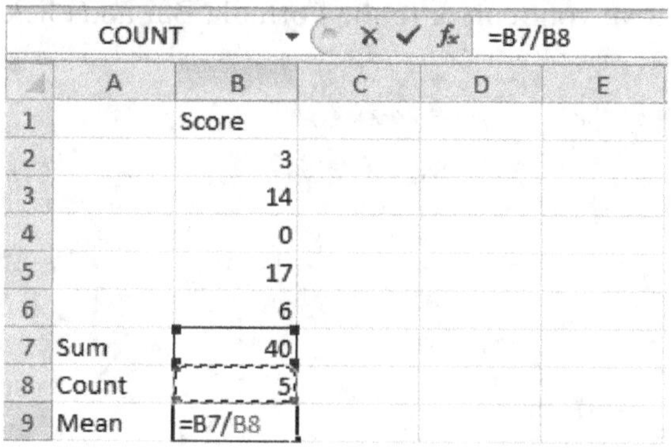

COUNT		▼	X ✓ fx	=B7/B8	
	A	B	C	D	E
1		Score			
2		3			
3		14			
4		0			
5		17			
6		6			
7	Sum	40			
8	Count	5			
9	Mean	=B7/B8			

13. Press [**Enter**]. You should now see a value of 8 in cell B9.

You could also type the cell addresses to enter them in formulas, but you will find that it is more accurate and faster to click in the cells.

	A	B
1		Score
2		3
3		14
4		0
5		17
6		6
7	Sum	40
8	Count	5
9	Mean	8

Deviation Scores

1. Activate cell **C1** and enter the label **Dev Score**.

2. Activate **C2**. To calculate a deviation score, we subtract the mean from the raw score. For the first score in the data set, this formula would be expressed as =B2-B9. Type = in cell C2. Then click in cell **B2**. Type - and then click in cell **B9**. Press [**Enter**] and you see –5 in cell C2.

	A	B	C
1		Score	Dev Score
2		3	=B2-B9
3		14	
4		0	
5		17	
6		6	
7	Sum	40	
8	Count	5	
9	Mean	8	

3. Dollar signs need to be added to B9 to make it an absolute reference. Click in **C2**. In the Formula Bar near the top of the screen, you should see =B2-B9.   Move your cursor to the Formula Bar and click and drag over **B9** so that it is highlighted.

COUNT		▾	× ✓ _fx_	=B2-B9	
	A	B	C	D	E
1		Score	Dev Score		
2		3	=B2-B9		

4. Press the **F4** key (usually located near the top left side of the keyboard). The cell reference should change to $B$9. Press [**Enter**]. Now, when you copy the formula in cell C2, each score will be divided by the value in cell B9.

The dollar signs could also be entered manually.

	COUNT		▾	× ✓	fx	=B2-$B$9
	A	B	C	D	E	
1		Score	Dev Score			
2			3 =B2-$B$9			

5. Copy the formula in C2 to cells C3 through C6. Your worksheet should now look like the one shown below.

	A	B	C
1		Score	Dev Score
2		3	-5
3		14	6
4		0	-8
5		17	9
6		6	-2
7	Sum	40	
8	Count	5	
9	Mean	8	

Squared Deviation Scores

1. Next, we will square each of the deviation scores. Activate cell **D1** and type **Sqd Dev Score**. Adjust the column width so that the entire label is displayed.

2. The formula that you will use to square the first deviation score is =C2^2. C2 is the cell address of the deviation score. The ^ is the exponentiation sign. Activate **D2**. Type =. Click in cell **C2**. Type **^2**. Press [**Enter**]. You will see 25 in cell D2.

	A	B	C	D
1		Score	Dev Score	Sqd Dev Score
2		3	-5	=C2^2
3		14	6	
4		0	-8	
5		17	9	
6		6	-2	
7	Sum	40		
8	Count	5		
9	Mean	8		

3. Copy cell D2 to cells D3 through D6. Your worksheet should now look like the one displayed below.

	A	B	C	D
1		Score	Dev Score	Sqd Dev Score
2		3	-5	25
3		14	6	36
4		0	-8	64
5		17	9	81
6		6	-2	4
7	Sum	40		
8	Count	5		
9	Mean	8		

Variance

1. Activate cell **A10**. Type **Variance**. Then activate cell **B10** where the value of the variance will be placed.

	A	B	C	D
1		Score	Dev Score	Sqd Dev Score
2		3	-5	25
3		14	6	36
4		0	-8	64
5		17	9	81
6		6	-2	4
7	Sum	40		
8	Count	5		
9	Mean	8		
10	Variance			

2. To calculate the variance, we will use the formula for a population variance. In this formula, the sum of the squared deviation scores is divided by the number of observations. For our example, that formula is expressed as =SUM(D2:D6)/B8. To begin the formula, click **AutoSum** located in the Editing group of the Home ribbon.   Because Excel automatically enters the range of numerical values immediately above the active cell, you will probably see =SUM(B8:B9). You need to change this to D2:D6.

	A	B	C	D
1		Score	Dev Score	Sqd Dev Score
2		3	-5	25
3		14	6	36
4		0	-8	64
5		17	9	81
6		6	-2	4
7	Sum	40		
8	Count	5		
9	Mean	8		
10	Variance	=SUM(B8:B9)		
11		SUM(number1, [number2], ...)		

3. Move the cursor to the Formula Bar. Click and drag over **B8:B9** so that it is highlighted.

COUNT			× ✓ fx	=SUM(B8:B9)
			SUM(**number1**, [number2], ...)	

	A	B	C
1		Score	Dev Score Sqd Dev Score
2		3	-5 25
3		14	6 36
4		0	-8 64
5		17	9 81
6		6	-2 4
7	Sum	40	
8	Count	5	
9	Mean	8	
10	Variance	=SUM(B8:B9)	

4. Next, in the worksheet, click and drag over **D2** through **D6**. You will see D2:D6 in both the Formula Bar and in cell B10.

COUNT			× ✓ fx	=SUM(D2:D6)
			SUM(**number1**, [number2], ...)	

	A	B	C
1		Score	Dev Score Sqd Dev Score
2		3	-5 25
3		14	6 36
4		0	-8 64
5		17	9 81
6		6	-2 4
7	Sum	40	
8	Count	5	
9	Mean	8	
10	Variance	=SUM(D2:D6)	

5. To complete the formula, in the Formula Bar, position the flashing I-beam at the end of the entries— after the second parenthesis.

6. Type **/** to indicate division. Then click in cell **B8**, the cell address of Count.

If you want to use the unbiased formula to estimate a population variance from sample data, the denominator is N-1. Edit the formula so that it reads =SUM(D2:D6)/(B8-1).

COUNT			× ✓ fx	=SUM(D2:D6)/B8

	A	B	C	D	E
1		Score	Dev Score	Sqd Dev Score	
2		3	-5	25	
3		14	6	36	
4		0	-8	64	
5		17	9	81	
6		6	-2	4	
7	Sum	40			
8	Count	5			
9	Mean	8			
10	Variance	=SUM(D2:D6)/B8			

7. Press [**Enter**]. A variance equal to 42 is now displayed in cell B10 of the worksheet.

◢	A	B	C	D
1		Score	Dev Score	Sqd Dev Score
2		3	-5	25
3		14	6	36
4		0	-8	64
5		17	9	81
6		6	-2	4
7	Sum	40		
8	Count	5		
9	Mean	8		
10	Variance	42		

Standard Deviation

1. Activate cell **A11**. Type **St Dev**. Then activate cell **B11** where the value of the standard deviation will be placed.

2. The standard deviation is equal to the square root of the variance. We will use Excel's square root function, SQRT. Type **=**. Then click the **Formulas** tab near the top of the screen and select **Insert Function**.

Instead of using Insert Function, you could type =SQRT. Both upper and lower case are accepted.

---

3. In the Insert Function dialog box, select the **Math & Trig** category and the **SQRT** function. Click **OK**.

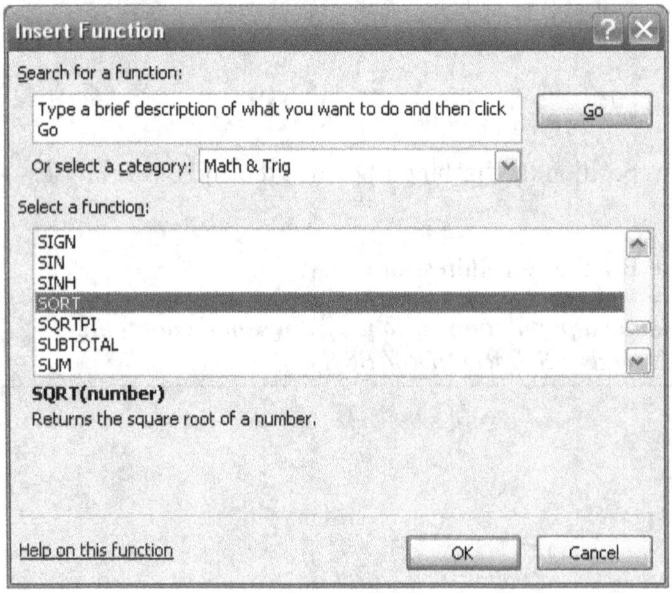

4. We want to take the square root of the variance which is located in cell B10. The flashing I-beam should be positioned in the Number window, indicating that it is ready for an entry. Click in cell **B10** of the worksheet.

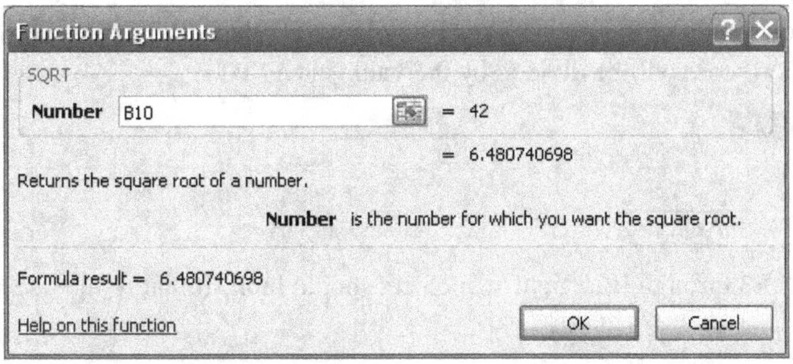

Function Arguments   ? X

SQRT

Number B10   = 42

                    = 6.480740698

Returns the square root of a number.

Number is the number for which you want the square root.

Formula result = 6.480740698

Help on this function   [OK]   [Cancel]

5. Click **OK**. A value of 6.480741 is now displayed in cell B11. If you increase the column width, you will see more decimal places. Note that 6.480740698 is displayed in the lower left of the dialog box shown above. By using Excel, you will not have rounding errors like those that typically occur when you carry out calculations by hand.

	A	B	C	D
1		Score	Dev Score	Sqd Dev Score
2		3	-5	25
3		14	6	36
4		0	-8	64
5		17	9	81
6		6	-2	4
7	Sum	40		
8	Count	5		
9	Mean	8		
10	Variance	42		
11	St Dev	6.480741		

z-Scores

1. As a last step, we will express each student's score as a *z*-score. Activate cell **E1** and type **z-score**.

2. Activate **E2** where the first *z*-score will be placed. The *z*-score is equal to the student's deviation score (in cell C2) divided by the standard deviation (in cell B11). In the Excel worksheet, the formula is expressed as =C2/\$B\$11. Type =. Click in cell **C2**. Type /. Click in cell **B11**. Press [**Enter**].

	A	B	C	D	E
1		Score	Dev Score	Sqd Dev Score	z-score
2		3	-5	25	=C2/B11

3. Dollar signs need to be added to B11 to make it an absolute reference. Click in **E2**. In the Formula Bar near the top of the screen, you should see =C2/B11. Move your cursor to the Formula Bar and click and drag over **B11** so that it is highlighted.

SQRT   ▼ × ✓ *fx* =C2/B11

	A	B	C	D	E
1		Score	Dev Score	Sqd Dev Score	z-score
2		3	-5	25	=C2/B11

4. Press the **F4** key. The cell reference should change to $B$11. Press [**Enter**]. Now, when you copy the formula in cell E2, each deviation score will be divided by the value in cell B11.

SQRT		▼	× ✓ fx	=C2/$B$11

	A	B	C	D	E
1		Score	Dev Score	Sqd Dev Score	z-score
2		3	-5	25	?/$B$11

5. Copy the formula in cell E2 to cells E3 through E6. Your worksheet should look similar to the one displayed below.

	A	B	C	D	E
1		Score	Dev Score	Sqd Dev Score	z-score
2		3	-5	25	-0.77152
3		14	6	36	0.92582
4		0	-8	64	-1.23443
5		17	9	81	1.38873
6		6	-2	4	-0.30861
7	Sum	40			
8	Count	5			
9	Mean	8			
10	Variance	42			
11	St Dev	6.480741			

6. Let's display the z-scores in column E with four decimal places. Click and drag over cells E2 through E6 so that they are highlighted.

	A	B	C	D	E
1		Score	Dev Score	Sqd Dev Score	z-score
2		3	-5	25	-0.77152
3		14	6	36	0.92582
4		0	-8	64	-1.23443
5		17	9	81	1.38873
6		6	-2	4	-0.30861
7	Sum	40			
8	Count	5			
9	Mean	8			
10	Variance	42			
11	St Dev	6.480741			

7. Click the **Home** tab near the top of the screen. Click the arrow in the Number group to decrease the decimal places until only four decimal places are displayed.

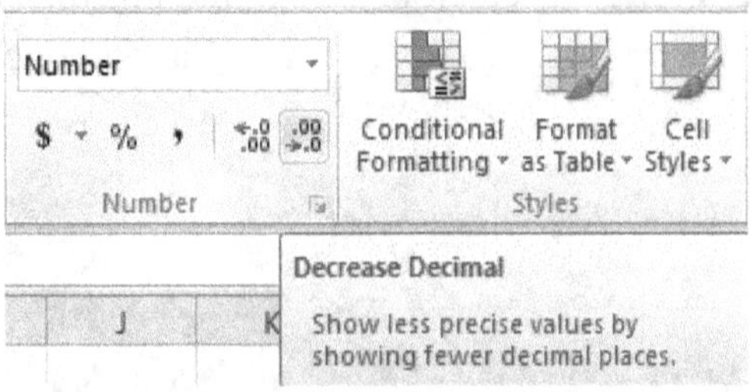

Number	▼
$ ▾ % , ◂.0 .00 .00 ▸.0	Conditional Format Cell Formatting ▾ as Table ▾ Styles ▾
Number	Styles

Decrease Decimal

Show less precise values by showing fewer decimal places.

	J	K

The completed worksheet is shown at the top of the next page.

	A	B	C	D	E
1		Score	Dev Score	Sqd Dev Score	z-score
2		3	-5	25	-0.7715
3		14	6	36	0.9258
4		0	-8	64	-1.2344
5		17	9	81	1.3887
6		6	-2	4	-0.3086
7	Sum	40			
8	Count	5			
9	Mean	8			
10	Variance	42			
11	St Dev	6.480741			

Because cell addresses were used in the formulas, corrections are easily made. For example, if you change the value in cell B2 from 3 to 2, you will see that all the other values will change accordingly.

8. You created a template in steps 1 through 7 that can now be used for any set of numbers. Let's use the scores **5, 16, 2, 19**, and **8**. In other words, we are adding a constant of 2 points to the scores that now appear in the worksheet. Enter these new values in column B as shown below.

	A	B	C	D	E
1		Score	Dev Score	Sqd Dev Score	z-score
2		5	-5	25	-0.7715
3		16	6	36	0.9258
4		2	-8	64	-1.2344
5		19	9	81	1.3887
6		8	-2	4	-0.3086
7	Sum	50			
8	Count	5			
9	Mean	10			
10	Variance	42			
11	St Dev	6.480741			

9. You will note that, when you add a constant, the sum and mean change, but all the other values stay the same. Let's try a different set of scores: 5, 4, 3, 2, 1.

	A	B	C	D	E
1		Score	Dev Score	Sqd Dev Score	z-score
2		5	2	4	1.4142
3		4	1	1	0.7071
4		3	0	0	0.0000
5		2	-1	1	-0.7071
6		1	-2	4	-1.4142
7	Sum	15			
8	Count	5			
9	Mean	3			
10	Variance	2			
11	St Dev	1.414214			

10. Now, all the values, except for count, have changed. You can also use this template for a different number of scores. All you have to do is insert or delete rows. For example, click in cell **A7**. **Right-click** and select **Insert** from the menu that appears.

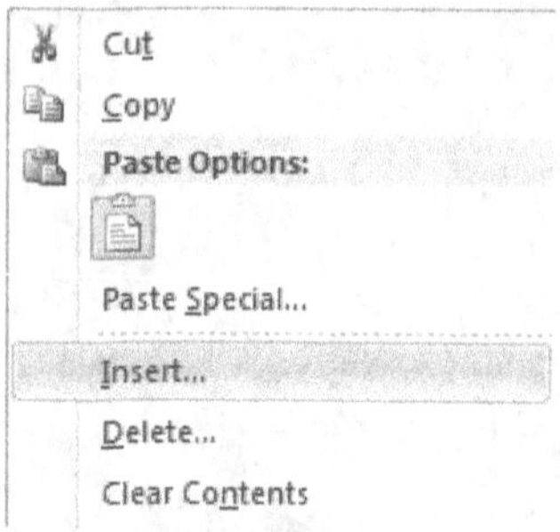

11.   Select **Entire row** and click **OK**.

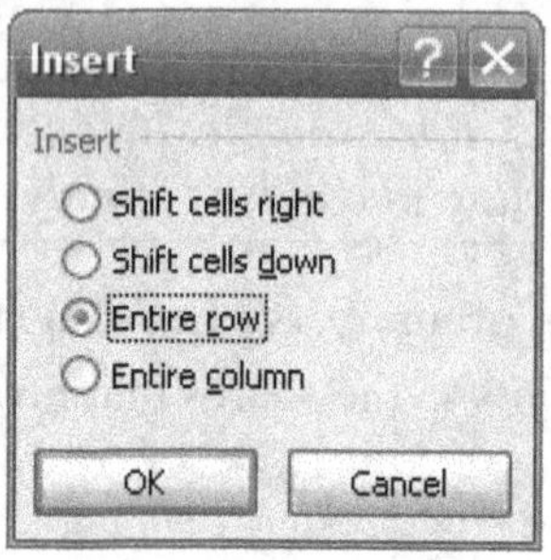

12.   A blank row 7 has been inserted. Enter these six scores in column B: **6, 5, 4, 3, 2, 1**. Because Excel copied the formulas into the inserted row, the calculations for all six scores were carried out. The completed worksheet is displayed below.

	A	B	C	D	E
1		Score	Dev Score	Sqd Dev Score	z-score
2		6	2.5	6.25	1.8257
3		5	1.5	2.25	1.0954
4		4	0.5	0.25	0.3651
5		3	-0.5	0.25	-0.3651
6		2	-1.5	2.25	-1.0954
7		1	-2.5	6.25	-1.8257
8	Sum	21			
9	Count	6			
10	Mean	3.5			
11	Variance	1.875			
12	St Dev	1.369306			

Chapter 4

Frequency Distributions

Excel provides three means of creating a frequency distribution: Histogram Analysis Tool, FREQUENCY function, and Pivot Table. The Histogram Analysis Tool and the FREQUENCY function can only be used with quantitative (e.g., age in years) variables. Although Pivot Table can be used to summarize both quantitative and qualitative (e.g., gender) variables, I recommend that you use the Histogram Analysis Tool for quantitative data when you want to create a graph.

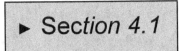 **Section 4.1**

Frequency Distributions Using Pivot Table and Pivot Chart

Sample Research Problem

A university instructor gathered information on students enrolled in an introductory world religions course. The instructor wanted to use this information to assign students to groups for class discussion and for semester projects. We will work with the data set, shown below, that contains information on 16 students.

	A	B	C
1	Gender	Age	Religion
2	M	19	Catholic
3	M	20	Jewish
4	F	22	None
5	M	23	Muslim
6	F	31	Catholic
7	F	22	Protestant
8	F	19	None
9	M	20	Jewish
10	M	19	Buddhist
11	F	20	Protestant
12	F	20	Protestant
13	M	19	Catholic
14	F	21	Protestant
15	M	19	Jewish
16	F	19	Catholic
17	F	20	Protestant

Frequency Distribution of a Quantitative Variable

I will use Age for my example of how to create a frequency distribution for a quantitative variable.

1. Open the "Ch4_Religions" worksheet on the Web site, or enter the student data shown above in an Excel worksheet.

2. Click in any cell containing data before you start the Pivot Table procedure so that the data range will be automatically entered in the Table/Range window. I clicked in cell A1.

3. Click the **Insert** tab near the top of the screen and select **Pivot Table**.

4. You should see Religions!$A$1:$C$17 in the Table/Range window of the Create Pivot Table dialog box. **Religions** is the name of the worksheet and **$A$1:$C$17** is the range that includes the data for all three variables. (If you prefer to enter the range manually, just type A1:C17 in the Table/Range window.) For this example, we will place the summary table in a new worksheet. To do this, just select **New Worksheet** in the lower part of the dialog box. Click **OK**.

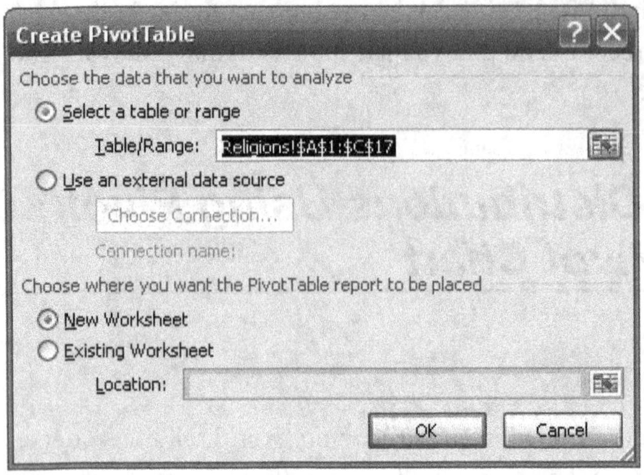

5. On the right side of the worksheet, you will see a Pivot Table Field List displaying Gender, Age, and Religion. These are the labels that appeared in the top cells (A1, B1, and C1) of the three columns of data. You want a summary for the Age variable, so click in the box to the left of Age to place a check mark there.

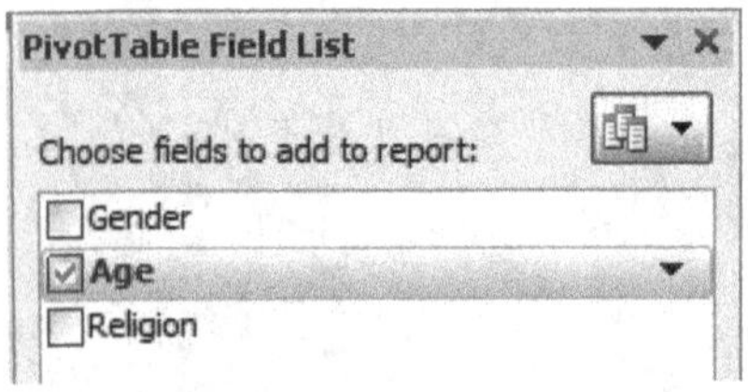

6. On the left side of the worksheet, you will see Sum of Age and 333. This is the sum of all the age values in the data set. You don't want this sum; you want the count of each age value. So that your display will look like the one shown in this manual, first click the down arrow under Pivot Table Field List at the right and select **Fields Section and Areas Section Stacked**.

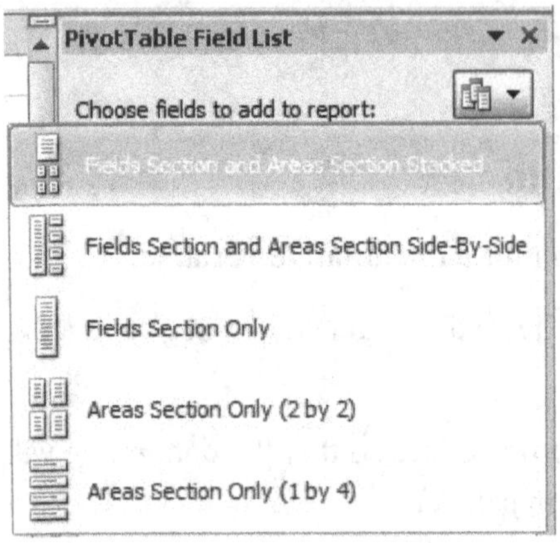

7. Next, click directly on **Age** in the field list and drag it down to the **Row Labels** section.

8. Click the down arrow to the right of **Sum of Age** and select **Value Field Settings**.

9. Select **Count** in the lower section of the Value Field Settings dialog box. Click **OK**.

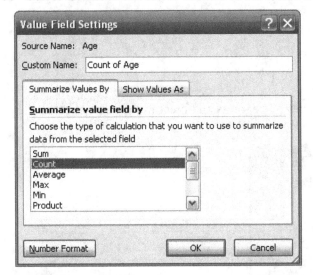

10. Your frequency distribution table should appear similar to the one shown at the top of the next page. In this Pivot Table report, **Grand Total** refers to the total number of observations in the frequency distribution. Our grand total is 16—the 16 students enrolled in the world religions course.

3	Row Labels	▾	Count of Age
4	19		6
5	20		5
6	21		1
7	22		2
8	23		1
9	31		1
10	**Grand Total**		**16**

11. It's very easy to create a histogram from the data in the Pivot Table report. First, click in any cell within the table. Then click the **Insert** tab at the top of the screen and select **Column** in the Charts group to insert a column chart.

12. Select the leftmost diagram in the 2-D Column row.

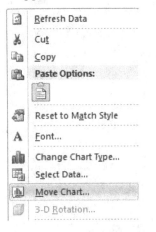

13. **Right-click** within the chart near the border and select **Move Chart** from the short cut menu that appears.

14. You are given the option of placing the chart in a new sheet or as an object in the same sheet as the table. For this example, select **New sheet**. Click **OK**.

15. Close the Pivot Table Field List on the right. The Design ribbon at the top includes a Chart Styles section that allows you to select color and styles. If you don't see a style that you like, click the lowest arrow on the right of the Chart Styles section and more color and style choices will appear. The Design

ribbon also includes a Chart Layouts section with options for titles, axis labels, and legends. Rather than use Chart Layouts, however, we will use labeling options available in the Layout ribbon.

The Design and Layout tabs are displayed only when the chart is active. To make the chart active, click anywhere within the chart so that you see small circles, squares, and/or dots along the border.

---

16. Let's start with a title for the chart. Click on the word **Total** so that it appears in a box. Delete **Total** and type **Age of Students Enrolled in World Religions**.

17. Next, let's add axis titles. Click the **Layout** tab at the top of the screen. Click **Axis Titles**. Select **Primary Horizontal Axis Title** and **Title Below Axis**.

18. Type **Age in Years** and press [**Enter**].

19. Click **Axis Titles** again. This time select **Primary Vertical Axis Title** and **Rotated Title**.

20. Type **Frequency** and press [**Enter**].

21. Let's delete the legend. To do this, click **Legend** and select **None**.

22. If you would like the frequency values displayed on the vertical bars of the chart, you can do this by selecting Data Labels where you will be given a number of options for placing the values. Click **Data Labels**, and for this example, select **Center**.

23. To change the font style or size, just right-click on the title or a value in the axis you would like to change and select Font from the menu that appears.

24. To change the amount of space between the bars, **right-click** on one of the vertical bars and select **Format Data Series** from the shortcut menu that appears.

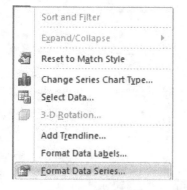

25. Click on the Gap Width arrow and drag it to the left to 50%. Click **Close** at the bottom of the dialog box.

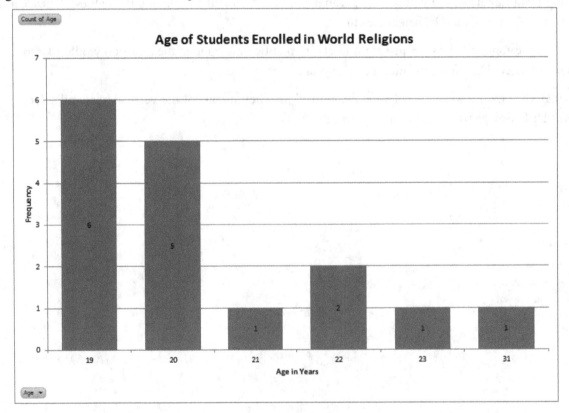

Your completed chart should look similar to the one shown below. Although this is an attractive histogram, you will notice that there are no values on the horizontal axis between 23 and 31 where you would most likely want to show the zero frequencies associated with 24 through 30. The histogram procedure that is available in Data Analysis Tools will show zero frequencies for these values. Directions for the histogram procedure are given in Section 4.2 of this chapter.

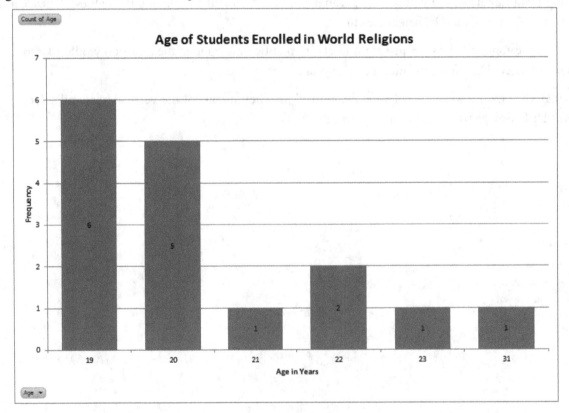

Frequency Distribution of a Qualitative Variable

I will use Religion to illustrate how to create a frequency distribution of a qualitative variable.

1. Open the "Ch4_Religions" worksheet on the Web site, or enter the student data shown on the first page of this chapter in an Excel worksheet. You will want to start the procedure on the worksheet where the Gender, Age, and Religion variables are displayed. If you just finished constructing a chart with the age data, the chart is most likely displayed rather than the data. To return to the data, just click the **Religions** sheet tab near the bottom of the screen.

2. Click in any cell containing data before you begin the Pivot Table procedure so that the data range will be automatically entered in the Table/Range window.

3. Click the **Insert** tab and select **Pivot Table** in the Insert ribbon. You are given the options of Pivot Table or Pivot Chart. Select **Pivot Table**.

4. You should see Religions!$A$1:$C$17 in the Table/Range window of the Create Pivot Table dialog box. For this example, we will place the summary table in a new worksheet. To do this, select **New Worksheet** in the lower part of the dialog box. Click **OK**.

5. On the right side of the worksheet, you will see a Pivot Table Field List displaying Gender, Age, and Religion. These are the labels that appeared in the top cells (A1, B1, and C1) of the three columns in the data range. You want a summary for the Religion variable, so click in the box to the left of Religion to place a check mark there.

6. On the left side of the worksheet, you will see an alphabetical listing of the Religions as shown at the top of the next page. You want the count for each religion. To obtain these counts, you want Religion displayed in the Row Labels of the Fields Section of the layout. If the Fields Section is not displayed,

click the down arrow under Pivot Table Field List at the right and select **Fields Section and Areas Section Stacked**.

3	Row Labels ▾
4	Buddhist
5	Catholic
6	Jewish
7	Muslim
8	None
9	Protestant
10	Grand Total

7. Then click directly on **Religion** and drag it down to the **Row Labels** section. (If you completed the Age chart just before beginning to work on this chart, Religion is probably already displayed in the Row Labels section as shown below.)

8. Click directly on **Religion** again and this time drag it down to the **Σ Values** section. You should see **Count of Religion** displayed in the window.

9. Your frequency distribution table should appear similar to the one shown at the top of the next page. In this Pivot Table report, you can easily see that the class membership includes 1 Buddhist, 4 Catholics, etc., and that the total enrollment is 16 students.

3	Row Labels	▼	Count of Religion
4	Buddhist		1
5	Catholic		4
6	Jewish		3
7	Muslim		1
8	None		2
9	Protestant		5
10	**Grand Total**		**16**

10.   We will now construct a bar chart of these data. First, click in any cell within the table. Then click the **Insert** tab at the top of the screen and select **Bar** in the Charts group to insert a bar chart.

11.   Select the leftmost diagram in the 2-D Bar row.

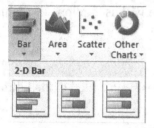

12.   **Right-click** within the chart somewhere near the border and select **Move Chart** from the shortcut menu that appears.

13.   You are given the option of placing the chart in a new sheet or as an object in the same sheet as the table. For this example, select **New sheet**. Click **OK**.

Your worksheet might automatically be given the name Chart2 because the age histogram constructed from the same data was given the name Chart1. If you would like, you can give a chart a name that more easily allows you to identify it. For example, you could name this chart "Gender graph" or just "Gender."

Move Chart   ? ✕

Choose where you want the chart to be placed:

⦿ New sheet: Chart1

○ Object in: Sheet1   ▾

                                              OK   Cancel

14. Close the Pivot Table Field List on the right. When the chart is active, you will see Design and Layout tabs at the top of the screen. If the chart is not active, just click anywhere on the chart. The dots, squares, and/or circles along the border indicate that the chart is active. We will use options available in the Design ribbon to add a main title and a legend.

15. Click the **Design** tab at the top of the screen.

16. Select **Layout 3** in the Chart Layouts section. You will see that a legend has been added at the bottom of the chart.

Chart Layouts

Layout 3

17. Let's modify the title at the top of the chart. Click on the word **Total** so that it appears in a box. Delete **Total** and type **Religion of Students Enrolled in World Religions**. Your completed bar chart should look similar to the one displayed below.

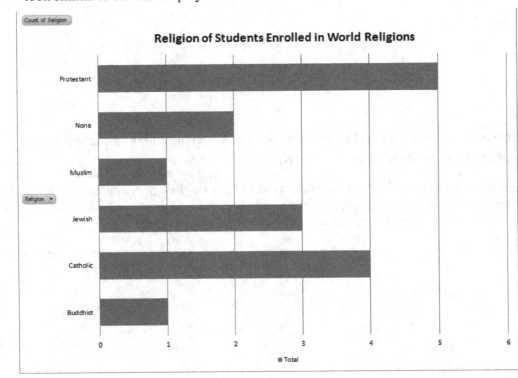

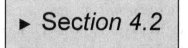

Frequency Distributions Using Data Analysis Tools

Histogram is one of the Data Analysis Tools. It provides a fairly quick means of producing both a frequency distribution table and a histogram of quantitative data. The Histogram Tool can generate a table and a graph of both ungrouped and grouped data. Both of these types will be illustrated.

Sample Research Problem

A researcher developed a test that was designed to measure cultural literacy. It consisted of multiple-choice items that asked about such things as historical events, science, art, literature, and the like. The maximum score on the test was 40.

Steps to Follow to Create a Frequency Distribution and a Histogram for Grouped Data

1. Open the "Ch4_Cultural Literacy" worksheet on the Web site, or enter the scores shown below in column A of an Excel worksheet.

	A	B
1	Cultural Literacy	
2	24	
3	35	
4	26	
5	29	
6	17	
7	22	
8	21	
9	17	
10	29	
11	26	
12	26	
13	16	
14	24	
15	22	
16	19	
17	25	
18	24	
19	23	
20	31	
21	14	
22	21	
23	23	

2. Excel's histogram procedure groups data and then displays it in a frequency distribution table and a frequency histogram. The procedure requires that you indicate a "bin" for each class. The number that you specify for each bin is actually the upper limit of the class. For the cultural literacy data, let's start with an upper limit of 15 and increase in increments of 3. Because 35 is the maximum score in the data set, the bins will end at 36. Enter **Bin** in cell C1 and key in the bin values as shown below.

	A	B	C
1	Cultural Literacy		Bin
2	24		15
3	35		18
4	26		21
5	29		24
6	17		27
7	22		30
8	21		33
9	17		36

3. Click the **Data** tab at the top of the screen and select **Data Analysis** in the Analysis Group.

If Data Analysis does not appear as a choice in the Data ribbon, you will need to load the Microsoft Excel ToolPak add-in. Follow the procedure on page 9.

4. In the Data Analysis dialog box, select **Histogram** and click **OK**.

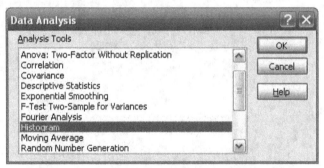

5. The Histogram dialog box will appear. Begin with the Input Range. This is the range of the Cultural Literacy scores. Click in the **Input Range** window of the Histogram dialog box. Then, in the worksheet, click and drag from cell A1 to cell A23. If you prefer, you can manually enter A1:A23 in the window.

The dialog box is often right on top of the data ranges that we want to mark for an analysis. You can move the dialog box by clicking on its title bar, holding down the mouse key, and dragging the box to a more convenient position.

6. Click in the **Bin Range** window. Then click and drag from cell C1 to cell C9. If you prefer, you can manually enter C1:C9 in the window.

7. Click in the box to the left of **Labels** to place a check mark there. The check mark lets Excel know that the first cell in each range contains a label rather than data. If there were no check mark, Excel would attempt to use the information in these cells when constructing the frequency distribution table and histogram. Because these cells contain words rather than data, however, you would receive an error message.

8. For **Output options, select New Worksheet Ply**. The output will be placed in a new worksheet with A1 as the uppermost left cell.

9. Click the **Chart Output** option so that a histogram will be produced along with the frequency distribution. The completed dialog box is shown below.

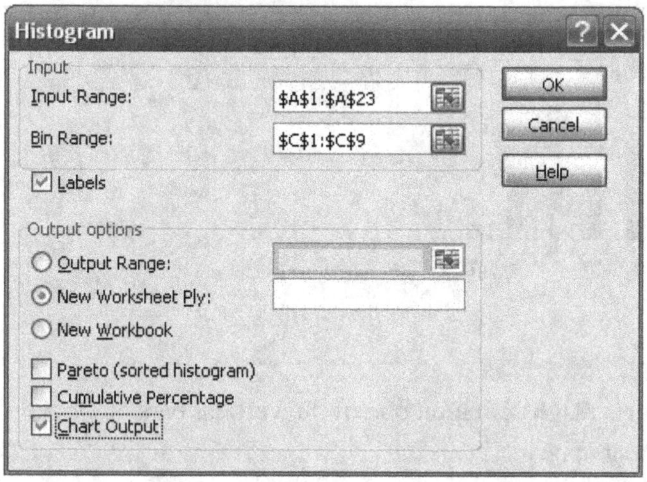

Click on Help (under OK and Cancel on the right side of the dialog box) to receive a detailed description of each required entry and option in the dialog box.

10. Click **OK**. You will receive output similar to that displayed below.

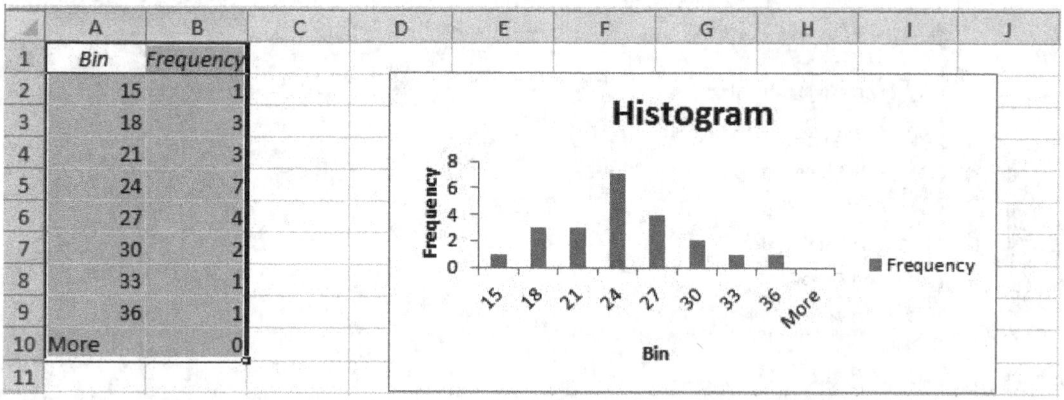

11. You will now follow steps to modify the histogram so that it is displayed in a more informative and attractive manner. First, make the chart taller so that it is easier to read. To do this, first click within the figure near a border. Small dots appear on the borders. Click on the dots on the bottom border of the figure and drag it down a few rows.

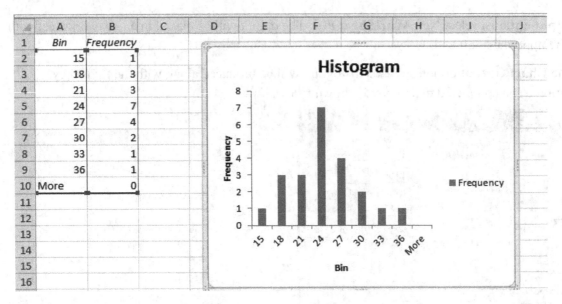

12.   Next remove the space between the vertical bars. **Right-click** on one of the vertical bars. Select
      Format Data Series from the shortcut menu that appears.

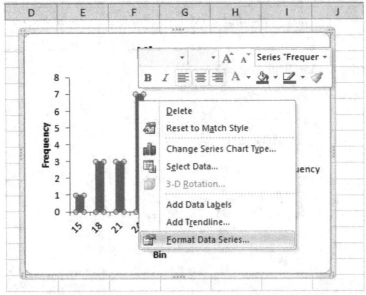

13.   Click on the Gap Width arrow and drag it all the way to the left to No Gap. Click **Close**.

14. Next, you will change the X-axis values from upper limits to ranges (e.g., 13-15, 16-18, etc.). You will be entering the ranges in column C of the Excel worksheet. But before you start, you need to select a Text format for the cells. First, click and drag over cells **C2:C9** so that they are highlighted. Then **right-click** and select **Format Cells** from the menu that appears.

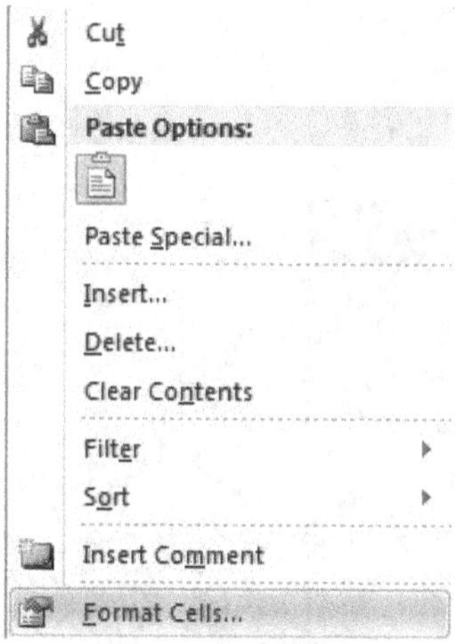

✂	Cu_t_
📋	_C_opy
📋	**Paste Options:**
	📋
	Paste _S_pecial...
	_I_nsert...
	_D_elete...
	Clear Co_n_tents
	Fil_t_er ▸
	S_o_rt ▸
	Insert Co_m_ment
📋	_F_ormat Cells...

15. Select **Text** and click **OK**.

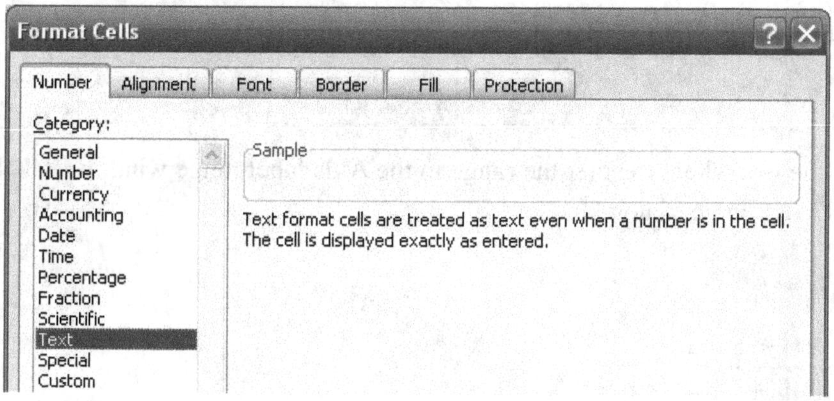

Format Cells   ? ✕

Number	Alignment	Font	Border	Fill	Protection

_C_ategory:

General
Number
Currency
Accounting
Date
Time
Percentage
Fraction
Scientific
Text
Special
Custom

Sample

Text format cells are treated as text even when a number is in the cell. The cell is displayed exactly as entered.

16. Enter the ranges in column C as shown below.

	A	B	C
1	Bin	Frequency	
2	15	1	13-15
3	18	3	16-18
4	21	3	19-21
5	24	7	22-24
6	27	4	25-27
7	30	2	28-30
8	33	1	31-33
9	36	1	34-36
10	More	0	

17. **Right-click** on a vertical bar in the chart. Select **Select Data** from the shortcut menu that appears.

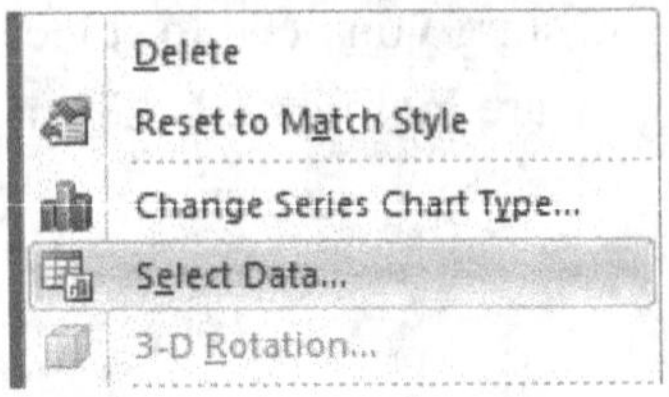

18. Click **Edit** under Horizontal (Category) Axis Labels. Click **OK**.

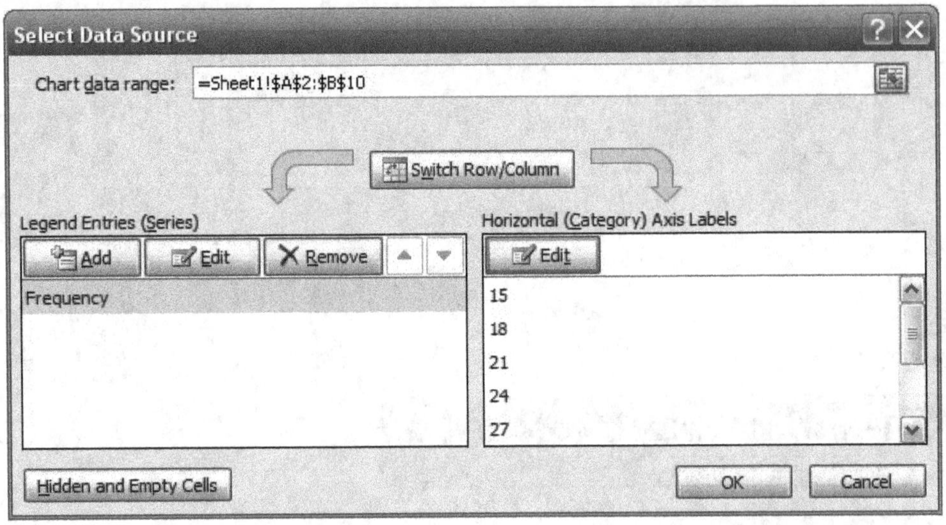

19. Click and drag over C2:C9 in the worksheet to enter the range in the Axis label range window. Click **OK**. Click **OK** in the Select Data Source dialog box.

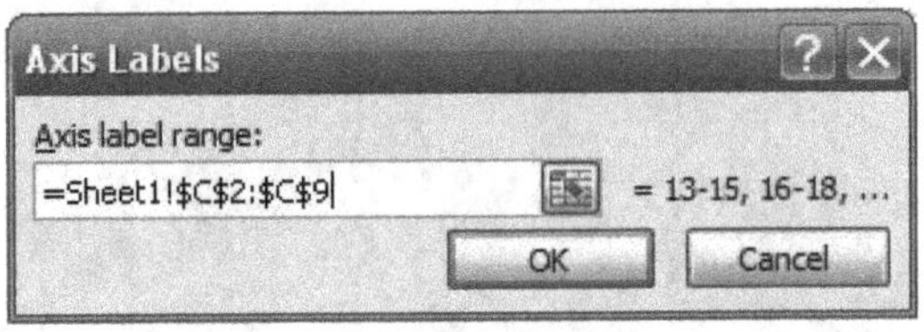

My frequency distribution and graph output is on Sheet 1 and that is why Sheet 1 appears in the range. Your output might be on a different sheet.

---

20. Let's move the histogram to a new worksheet before we change the titles. To move the histogram, **right-click** in the blank space around the histogram near a border. Select **Move Chart** from the menu.

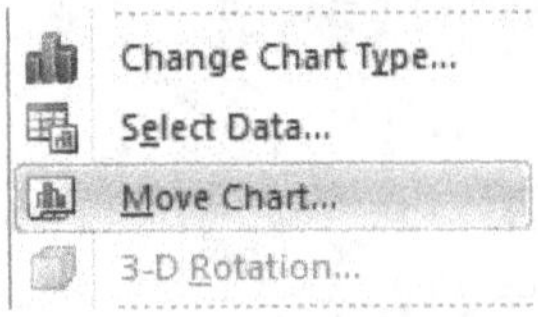

21. Select **New sheet** and click **OK**.

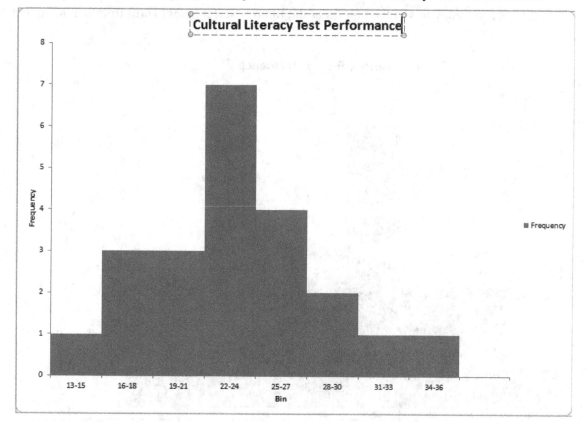

22. Let's change the title at the top of the histogram. Click directly on **Histogram** and a box will appear around it. Delete **Histogram** and replace it with **Cultural Literacy Test Performance**.

23. Next, let's change the X-axis title. Click directly on **Bin** at the bottom of the figure. Delete **Bin** and replace it with **Test Score**.

24. Let's remove the frequency legend at the right. Click the **Layout** tab at the top of the screen. Click **Legend** in the Labels group and select **None**.

25. Finally, let's display the frequency values by adding a data table. Select **Data Table** in the Layout ribbon.   Select **Show Data Table with Legend Keys**.

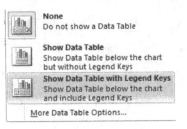

None
Do not show a Data Table

Show Data Table
Show Data Table below the chart but without Legend Keys

Show Data Table with Legend Keys
Show Data Table below the chart and include Legend Keys

More Data Table Options...

26. The rightmost frequency value is a zero. This zero value is displayed because of the "More" category that appears at the bottom of the bin list in the frequency table. Let's get rid of this zero. **Right-click** on any value in the Frequency row at the bottom of the figure. Select **Select Data** from the menu that appears.

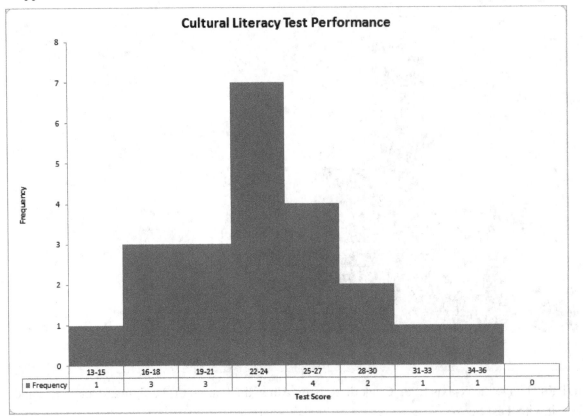

Cultural Literacy Test Performance

	13-15	16-18	19-21	22-24	25-27	28-30	31-33	34-36	
▥ Frequency	1	3	3	7	4	2	1	1	0

Test Score

27.   Click **Edit** under Legend Entries (Series) on the left.

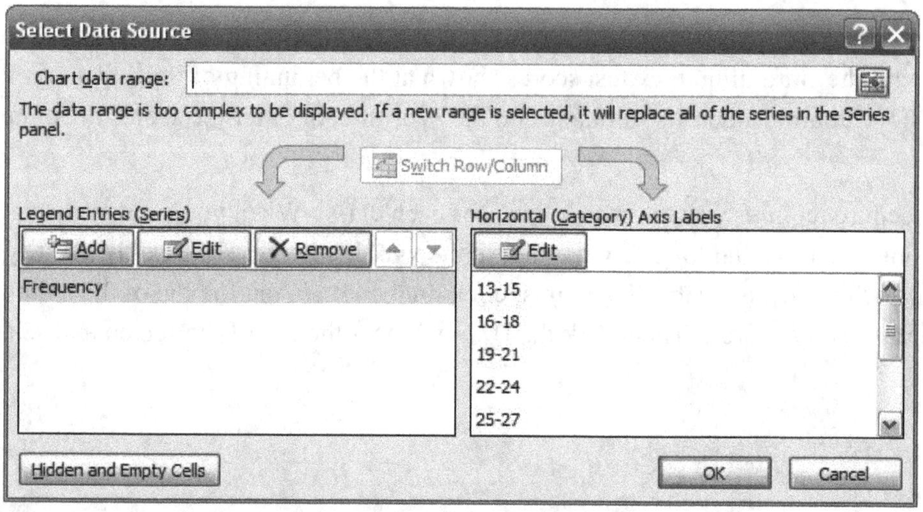

Select Data Source

Chart data range:

The data range is too complex to be displayed. If a new range is selected, it will replace all of the series in the Series panel.

Switch Row/Column

Legend Entries (Series)

Add   Edit   Remove   ▲   ▼

Frequency

Horizontal (Category) Axis Labels

Edit

13-15
16-18
19-21
22-24
25-27

Hidden and Empty Cells   OK   Cancel

28.   The zero frequency appeared in cell B10 of the frequency table. So change the last number in the Series values window from 10 to 9 as shown below. Click **OK**. Click **OK** in the Select Data Source dialog box.

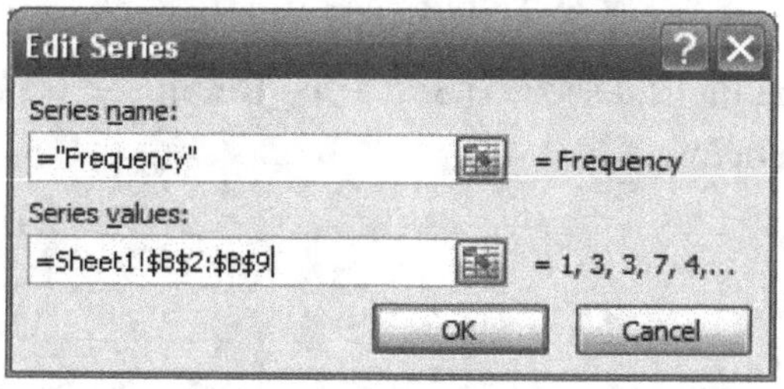

Edit Series

Series name:
="Frequency"   = Frequency

Series values:
=Sheet1!$B$2:$B$9   = 1, 3, 3, 7, 4,...

OK   Cancel

Your completed histogram should look similar to the one shown here.

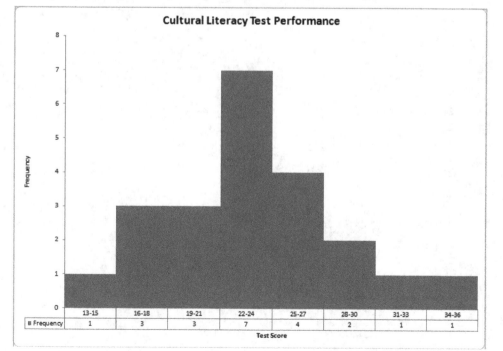

Cultural Literacy Test Performance

	13-15	16-18	19-21	22-24	25-27	28-30	31-33	34-36
■ Frequency	1	3	3	7	4	2	1	1

Test Score

Steps to Follow to Create a Frequency Distribution and a Histogram for Ungrouped Data

1. You will be working with the cultural literacy test scores shown at the beginning of Section 4.2 in this chapter. Open the "Ch4_Cultural Literacy" worksheet on the Web site, or enter the scores in column A of an Excel worksheet.

2. Excel's histogram procedure requires you to enter a "bin" for each class. When you want to display ungrouped data, you simply enter a bin for every number in the variable range. To quickly identify the highest and lowest values, just sort the data. To do this, click on the letter A at the top of the column that contains the cultural literary scores. Then click the **Data** tab near the top of the screen and select the **A to Z** sort.

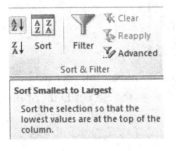

3. You can now easily see that the lowest score is 14 and the highest score is 35. Type **Bin** in cell C1 and enter the bin values from 14 to 35 in column C as shown below.

	A	B	C
1	Cultural Literacy		Bin
2	14		14
3	16		15
4	17		16
5	17		17
6	19		18
7	21		19
8	21		20
9	22		21
10	22		22
11	23		23
12	23		24
13	24		25
14	24		26
15	24		27
16	25		28
17	26		29
18	26		30
19	26		31
20	29		32
21	29		33
22	31		34
23	35		35

4. Select **Data Analysis** in the Data ribbon.

If Data Analysis does not appear as a choice in the Data ribbon, you will need to load the Microsoft Excel ToolPak add-in. Follow the procedure on page 9.

5. Select **Histogram** and click **OK**.

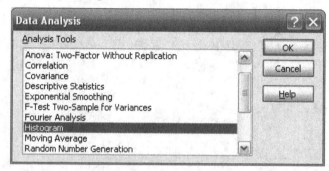

6. The Histogram dialog box will appear. Enter the **Input Range** and **Bin Range** as shown below. Also click the **Labels** box, the **New Worksheet Ply** button, and the **Chart Output** box.

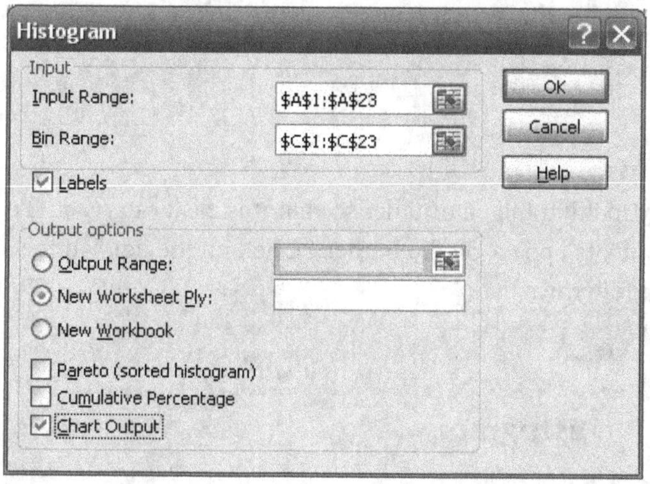

*Be sure to select **Chart Output** at the bottom of the Histogram dialog box. Otherwise you will get a frequency distribution table and no chart.*

7. Click **OK**. You will receive output similar to the output displayed at the top of the next page.

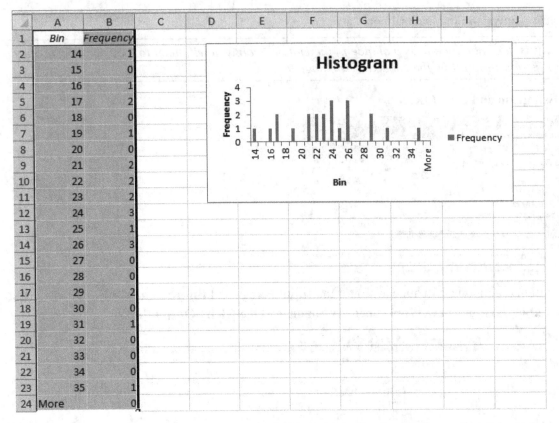

8. You will now modify the histogram. Begin by making the chart taller so that it is easier to read. To do this, first click within the figure near a border. Dots appear on the border. Click on the dots on the bottom border of the figure and drag it down a few rows.

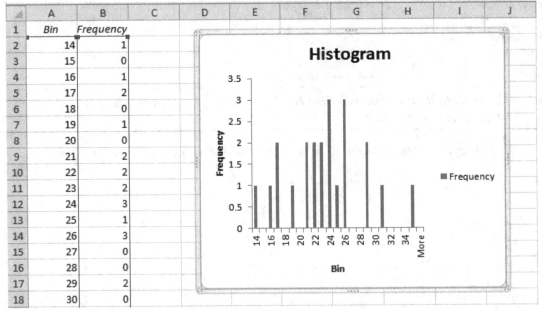

9. Next, remove the space between the vertical bars. **Right-click** on one of the vertical bars. Select **Format Data Series** from the shortcut menu that appears.

10. Click on the Gap Width arrow and drag it all the way to the left to No Gap. Click **Close**.

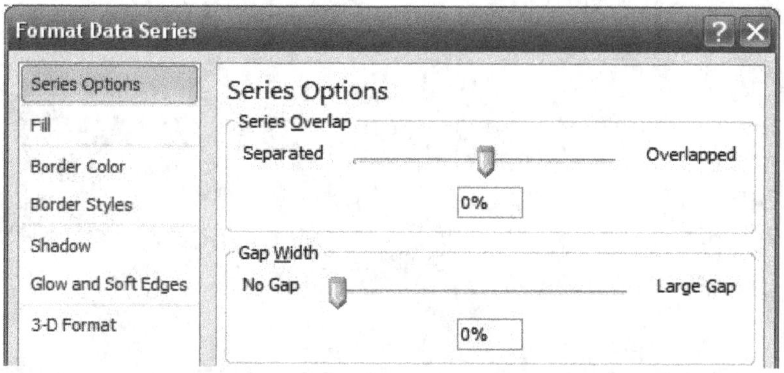

11. Next, you will make a number of modifications to the histogram. Let's start by getting rid of the word "More" on the right end of the X-axis. **Right-click** on any value in the X-axis. Select **Select Data** from the shortcut menu that appears.

12. Click **Edit** on the right side of the dialog box under Horizontal (Category) Axis Labels.

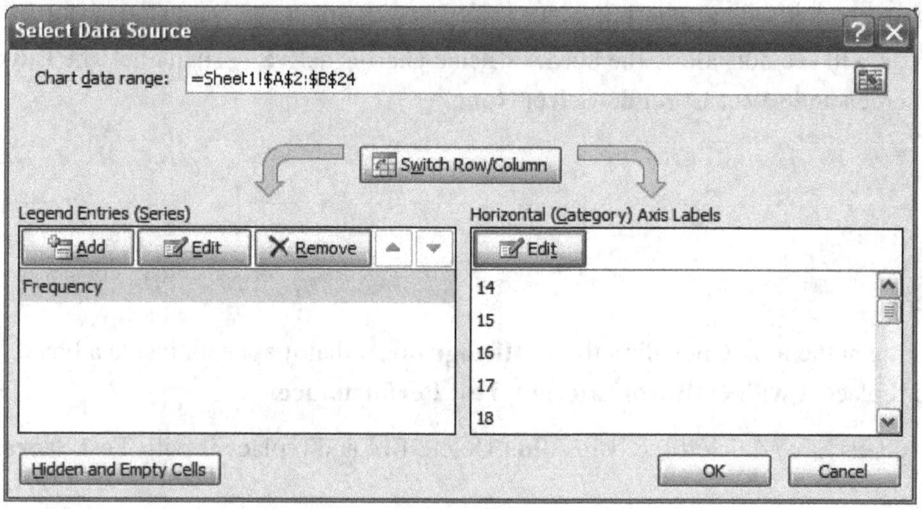

13. "More" is in cell A24. So change the last number in the axis label range from 24 to 23 as shown below. Click **OK**. Click **OK** in the Select Data Source dialog box.

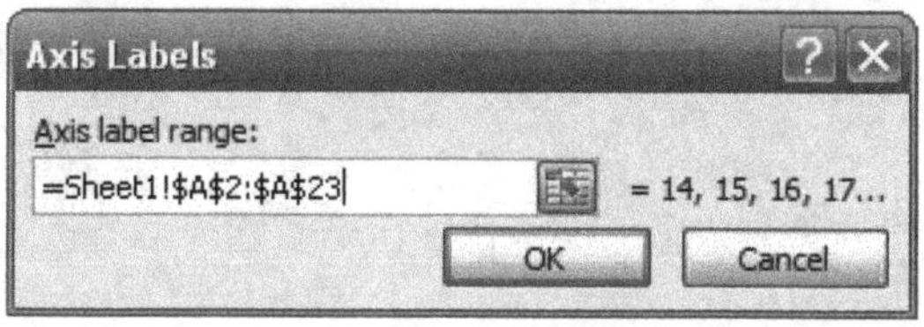

14. You will probably want to have whole numbers on the Y-axis scale rather than decimal values. To make this change, **right-click** on any value in the Y-axis scale and select **Format Axis** from the shortcut menu.

Delete

Reset to Match Style

Font...

Change Chart Type...

Select Data...

3-D Rotation...

Add Major Gridlines

Add Minor Gridlines

Format Axis...

15. Under Axis Options on the right side, click the **Fixed** button to the right of Major unit. Type **1** in the box. Click **Close**.

Format Axis   ? X

Axis Options

Axis Options	Axis Options			
Number	Minimum:	⦿ Auto	◯ Fixed	0.0
Fill	Maximum:	⦿ Auto	◯ Fixed	3.5
Line Color	Major unit:	◯ Auto	⦿ Fixed	1
Line Style	Minor unit:	⦿ Auto	◯ Fixed	0.1

16. Next, let's delete the Frequency legend at the right. Click anywhere on the histogram so that the histogram is active. (You will see dots along the border when a chart is active.) Then click the **Layout** tab near the top of the screen and select **Legend**. Select **None**.

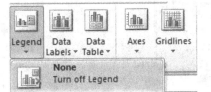

Legend   Data   Data   Axes   Gridlines
         Labels Table

None
Turn off Legend

17. Let's change the main title at the top. Click directly on **Histogram** so that it appears inside a box. Delete **Histogram** and replace it with **Cultural Literacy Test Performance**.

18. Next, let's change the X-axis title. Click directly on **Bin**. Delete **Bin** and replace it with **Test Score**.

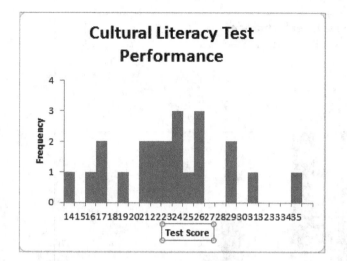

19. Next, let's add gridlines. Click the **Layout** tab near the top of the screen. Select **Gridlines**, **Primary Horizontal Gridlines**, and **Major Gridlines**.

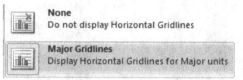

20. Finally, let's move the histogram to a new worksheet. **Right-click** in the blank space around the histogram near a border. Select **Move Chart** from the menu.

21. Select **New sheet** and click **OK**.

Your completed histogram should look similar to the one shown at the top of the next page.

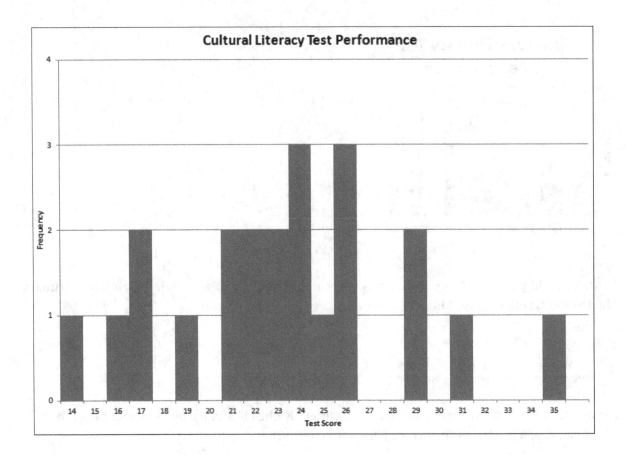

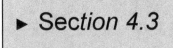 Section 4.3   ***Frequency Distributions Using the FREQUENCY Function***

The FREQUENCY function provides a quick means of obtaining a frequency distribution for an array of values. FREQUENCY returns the distribution as a vertical array of numbers. Because the result is returned as an array, the data must be entered as an array formula. For our example, we'll use the cultural literacy test scores shown at the beginning of Section 4.2 in this chapter.

1. Open the "Ch4_Cultural Literacy" worksheet on the Web site, or enter the scores in column A of an Excel worksheet.

2. We'll use the same bins as those used to illustrate a histogram for grouped data in Section 4.2. Type **Bin** in cell C1 and key in the bin values as shown at the top of the next page.

	A	B	C
1	Cultural Literacy	Bin	
2	24		15
3	35		18
4	26		21
5	29		24
6	17		27
7	22		30
8	21		33
9	17		36

3. We will display the output in column D. Type **Frequency** in cell D1 so that the column with the results has a label. Then click and drag over the range **D2:D10** to select these cells for the output. Because the output includes one more element than the number of bins, we need to select an output range that has one more cell than the bin range.

	A	B	C	D
1	Cultural Literacy	Bin		Frequency
2	24		15	
3	35		18	
4	26		21	
5	29		24	
6	17		27	
7	22		30	
8	21		33	
9	17		36	
10	29			

4. Click the **Formulas** tab near the top of the screen and select **Insert Function**.

5. In the Insert Function dialog box, select the **Statistical** category and the **FREQUENCY** function. Click **OK**.

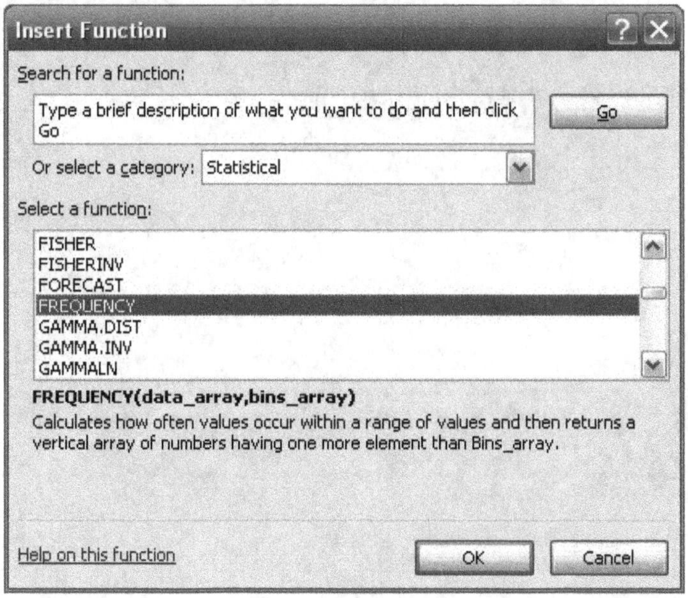

Insert Function

Search for a function:

Type a brief description of what you want to do and then click Go   Go

Or select a category: Statistical

Select a function:

FISHER
FISHERINV
FORECAST
FREQUENCY
GAMMA.DIST
GAMMA.INV
GAMMALN

FREQUENCY(data_array,bins_array)
Calculates how often values occur within a range of values and then returns a vertical array of numbers having one more element than Bins_array.

Help on this function   OK   Cancel

6. Fill in the FREQUENCY dialog box as shown below. The data array range is **A1:A23** and the bins array range is **C1:C9**. After you have entered these ranges, press **[CTRL]+[SHIFT]+[Enter]** so that the input is entered as arrays.

[CTRL]+[SHIFT]+[Enter] means that you should hold all three keys down at the same time.

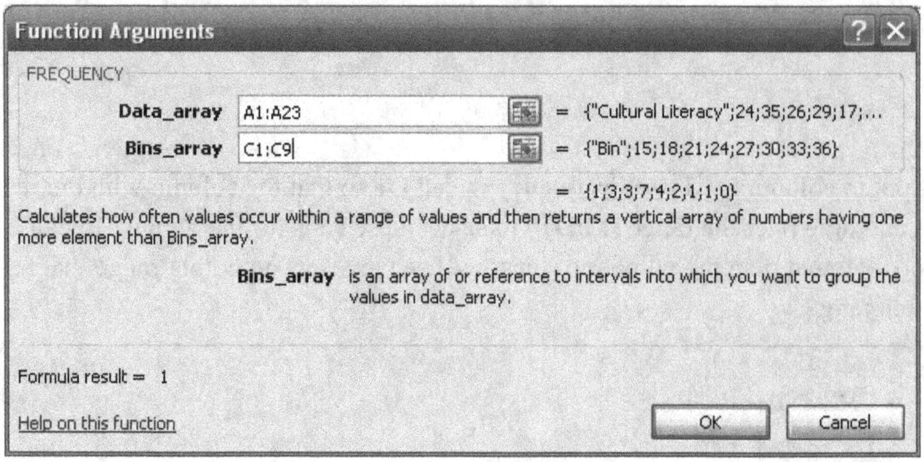

Function Arguments   ? X

FREQUENCY

Data_array	A1:A23	📷	= {"Cultural Literacy";24;35;26;29;17;...
Bins_array	C1:C9	📷	= {"Bin";15;18;21;24;27;30;33;36}

= {1;3;3;7;4;2;1;1;0}

Calculates how often values occur within a range of values and then returns a vertical array of numbers having one more element than Bins_array.

Bins_array   is an array of or reference to intervals into which you want to group the values in data_array.

Formula result =   1

Help on this function   [OK]   [Cancel]

Your output should look similar to the output shown below. Notice the set of braces around the function in the formula line. These braces indicate that the input was entered as an array formula. The frequency of 0 in cell D1 is associated with a bin that would be labeled "more" in Pivot Table or Histogram Tool output. For this example, the "more" bin contains any values in the data set that are higher than 36.

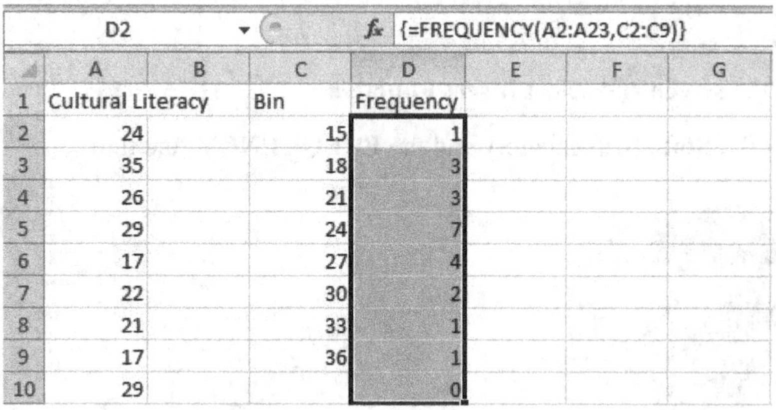

	D2	▼	*fx*	{=FREQUENCY(A2:A23,C2:C9)}			
	A	B	C	D	E	F	G
1	Cultural Literacy		Bin	Frequency			
2	24		15	1			
3	35		18	3			
4	26		21	3			
5	29		24	7			
6	17		27	4			
7	22		30	2			
8	21		33	1			
9	17		36	1			
10	29			0			

Excel provides a variety of ways to obtain summary measures for a set of data. I have included three of these in this chapter: Data Analysis Tools, Functions, and Pivot Table. I present Data Analysis Tools first because it is the simplest of the three. I end with the Pivot Table, which is easy to use but does not offer as many summary measures as Data Analysis Tools or Excel's Functions.

▶ Section 5.1	Data Analysis Tools: Descriptive Statistics

Data Analysis is located in the Data ribbon. The Descriptive Statistics Analysis Tool provides a quick and simple way of obtaining, simultaneously, several descriptive measures for a single variable or for two or more variables. These measures include: sum, count, mean, median, mode, minimum, maximum, range, standard deviation, variance, standard error, kurtosis, and skewness.

Sample Research Problem

An electric utility company wanted to gather information regarding energy usage in its service territory. The company conducted an extensive survey that included many questions, two of which were the number of television sets and microwave ovens owned by residential customers. Another question asked whether the customer lived in an apartment or a single-family house. I have used the responses provided by 11 customers to explain how to obtain descriptive statistics utilizing the Descriptive Statistics Analysis Tool.

Steps to Follow to Obtain Descriptive Statistics for One Variable

1. Open the "Ch5_Energy Use" worksheet on the Web site or enter the data as shown below in an Excel worksheet.

	A	B	C
1	TV	Microwave	Residence
2	3	1	Apt
3	0	1	House
4	9	2	Apt
5	2	1	House
6	3	3	House
7	1	0	House
8	2	1	Apt
9	0	2	Apt
10	3	1	House
11	5	1	House
12	1	1	Apt

2. Let's begin by obtaining descriptive statistics for TV ownership. Click the **Data** tab near the top of the screen and select **Data Analysis** in the Analysis group at the right.

If Data Analysis does not appear as a choice in the Data ribbon, you will need to load the Microsoft Excel ToolPak add-in. Follow the procedure on page 9.

---

3. A list of Analysis Tools is displayed in the Data Analysis dialog box. Select **Descriptive Statistics** and click **OK**.

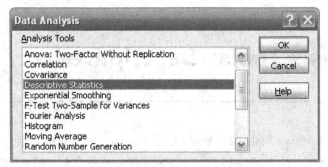

4. Complete the Descriptive Statistics dialog box as shown below. A description of each entry immediately follows the dialog box.

- **Input Range**. Enter the worksheet location of the data you wish to analyze. The TV data are located in A1:A12. Click in the **Input Range** window. Click in cell **A1** and then drag to cell **A12**. If you prefer, you can also key in the range by manually entering A1:A12. Notice that we have included A1 in the range. A1 contains "TV." TV is referred to as a label.
- **Grouped By**. Excel gives you the option of entering each variable in its own column or its own row. For this example, we used columns, which is the default option. There should be a black dot in the button to the left of **Columns**. If we had instead grouped our data by rows, we would have needed to click in the Rows button.
- **Labels in First Row**. I have found that it is very helpful to label output clearly. Therefore, I included a label in the first row for each variable (i.e., "TV," "Microwave," "Residence.") The labels will not be

included in the computations but will be used to label the output. (Note, however, that labels are required for the Pivot Table procedure that will be described later in this chapter.) Click in the box for **Labels in First Row**.

For this example, if you forget to click Labels in First Row, you will receive the following error message: **Input range contains nonnumeric data**. *If the labels are numeric, such as ages (e.g., 8, 9, 10) or years (e.g., 2009, 2010, 2011), and you forget to click Labels in First Row, you will not receive an error message. However, the numeric labels will be included in the computations and your output will be inaccurate. An indication that this has happened is seeing labels in your output such as* **Column 1** *and* **Column 2**.

- **Output options**. Three output options are available.

  Output Range. Select Output Range if you wish to place the output in the same worksheet as the data. Enter the upper left cell of the desired location of the output table in the Output Range window. For this example, let's place the output table next to the data. Click in the **Output Range** window and then click in cell **E1** of the worksheet. If you prefer, you can manually enter **E1** in the Output Range window.

  New Worksheet Ply. This is the default option. Unless specified otherwise, the output table will be pasted into a new worksheet with A1 as the upper left cell. If you like, you can give the worksheet a special name.

  New Workbook. Select New Workbook if you want to create a new workbook to display the analysis output. The analysis output will be pasted into a worksheet of a new workbook with A1 as the upper left cell of the output table.

- **Summary statistics**. Summary statistics provided by the Descriptive Statistics Analysis Tool include the mean, standard error of the mean, median, mode, standard deviation, variance, kurtosis, skewness, range, minimum, maximum, sum, and count. Click the **Summary statistics** box so that Excel will provide these values.

If you do not select summary statistics, confidence level for mean, Kth largest, or Kth smallest, you will receive this error message: **Please choose at least one statistical option**.

- **Confidence Level for Mean**. We would like output regarding the confidence level. So, click in the box next to **Confidence Level for Mean**. The default is 95%, which is what we will use for this example. If you would like a different confidence level (e.g., 90%), enter that level in the Confidence Level for Mean window. Any number greater than 0 and less than 100 can be entered.

- **Kth Largest**. In addition to the maximum value in a data set, we might also want to know, for example, the second largest or third largest value. If so, we would click in the box next to Kth Largest and then enter the desired value in the Kth Largest window. For this example, let's ask for the second largest value. Click in the **Kth Largest** box and enter **2** in the window. The default is 1, which is the maximum value.

- **Kth Smallest**. In addition to the minimum value in a data set, we might also be interested in knowing another small value such as the second smallest. To obtain this value, click in the box next to **Kth Smallest** and then enter **2** in the window. The default is 1, which is the minimum value.

5. Click **OK**, and the output table will be displayed with E1 as the upper left cell as requested. Adjust the column width as necessary to display all labels.

Interpreting the Output

The Descriptive Statistics output table is shown below in columns E and F. An explanation of each of the measures that is included in Descriptive Statistics is presented immediately after the output table.

	A	B	C	D	E	F
1	TV	Microwave	Residence		TV	
2	3	1	Apt			
3	0	1	House		Mean	2.636364
4	9	2	Apt		Standard Error	0.777791
5	2	1	House		Median	2
6	3	3	House		Mode	3
7	1	0	House		Standard Deviation	2.579641
8	2	1	Apt		Sample Variance	6.654545
9	0	2	Apt		Kurtosis	3.212809
10	3	1	House		Skewness	1.602855
11	5	1	House		Range	9
12	1	1	Apt		Minimum	0
13					Maximum	9
14					Sum	29
15					Count	11
16					Largest(2)	5
17					Smallest(2)	0
18					Confidence Level(95.0%)	1.733026

- **Mean**. The arithmetic average of the number of TVs owned by the 11 residential customers in the electric utility's sample, $\overline{X} = 2.6364$ TVs.

- **Standard Error**. Standard error of the mean, $S_{\overline{X}}$, calculated using the formula $\dfrac{S}{\sqrt{n}}$, where S is the sample standard deviation and n is the number of observations. For the TV ownership distribution, $S_{\overline{X}} = 0.7778$.

- **Median**. The observation that splits the distribution in half. To determine the value of the median, the observations are first arranged in either ascending or descending order. If the number of observations is even, the median is found by taking the arithmetic average of the two middle observations. If the number of observations is odd, then the median is the middle observation. When arranged in ascending order, our sample data would be: 0, 0, 1, 1, 2, 2, 3, 3, 3, 5, 9. Because the number of observations is odd $(n = 11)$, the median is the middle (6th) observation, which is equal to 2.

- **Mode**. Observation value associated with the highest frequency. For this example, the mode is equal to 3.

A word of caution is in order regarding the value reported for the mode. Three situations are possible regarding the mode: 1) If all values occur only once in a distribution, Excel will return #N/A.   2) If a variable has only one mode, Excel will return that value.   3) If a variable has more than one mode, however, Excel will still return only one value. The value that is returned will be the one associated with the modal value that occurs first in the data set. To check the accuracy of the mode, I suggest that you create a frequency distribution.

- **Standard Deviation**. Unbiased estimate of the population standard deviation found by applying the formula shown at the top of the next page.

$$S = \sqrt{\frac{\sum (X - \bar{X})^2}{n-1}}$$

For the TV ownership distribution, $S = 2.5796$.

- **Sample Variance.** Standard deviation squared (S^2). For the TV distribution, $S^2 = 6.6545$.

- **Kurtosis.** A numerical index that describes a distribution with respect to its flatness or peakedness as compared to a normal distribution. A negative value characterizes a relatively flat distribution, whereas a positive value characterizes a relatively peaked distribution. Note that a distribution must contain at least four observations in order for kurtosis to be calculated. Kurtosis for our sample problem is 3.2128, indicating that the TV distribution is somewhat peaked as compared to a normal distribution.

- **Skewness.** A numerical index that characterizes the asymmetry of a distribution. Negative skew indicates that the longer tail extends in the direction of low values in the distribution. Positive skew, in contrast, indicates that the longer tail extends in the direction of the high values in the distribution. Note that skewness cannot be computed when there are fewer than three observations in the distribution. The positive value of 1.6029 for our sample research problem indicates that the distribution of TV ownership is positively skewed. The positive skew is primarily due to the nine TVs owned by one of the survey respondents.

- **Range.** The minimum value in a distribution subtracted from the maximum value. The range for our distribution is found by subtracting 0 from 9, producing a range equal to 9.

- **Minimum.** The lowest value occurring in the distribution. The lowest number of TVs owned by customers in our sample is 0.

- **Maximum.** The highest value occurring in the distribution. The highest number of TVs owned by a customer in our sample is 9.

- **Sum.** Sum of the values in the distribution. For our sample research problem, $\sum X = 29$.

- **Count.** Number of observations in the distribution. For our sample research problem, $n = 11$.

- **Largest(2).** Second largest value in the distribution. When the data are arranged in descending order, the second largest value is the number that occurs right after the number in first position. The second largest number of TVs owned by customers in our sample is 5.

- **Smallest(2).** Second smallest value in the distribution. When the data are arranged in ascending order, the second smallest value is the number that occurs immediately after the number in first position. The second smallest number of TVs owned by customers in our sample is 0.

- **Confidence Level (95.0%).** Amount of error subtracted from and added to the sample mean when constructing a confidence interval for the population mean. To calculate the confidence level using the Descriptive Statistics Analysis Tool, it is assumed that the population variance is not known and is estimated by the sample variance. The confidence interval takes the general form shown below.

$$\bar{X} \pm (t_{Crit})(S_{\bar{X}})$$

In this formula, $\bar{X}$ serves as an estimate of the population mean, and the error component is calculated by multiplying the standard error of the mean times the two-tailed critical value of t for the selected value of alpha. The t critical value (2.2281) for our sample research problem is found in the t distribution with $n - 1 = 10$ degrees of freedom and alpha equal to .05. The standard error is equal to 0.7778. The product of 2.2281 and 0.7778 is 1.7330. The 95% confidence interval, therefore, is:

$$2.6364 - 1.7330 \leq \mu \leq 2.6364 + 1.7330$$

$$0.9034 \leq \mu \leq 4.3694$$

Steps to Follow to Obtain the Population Variance and Population Standard Deviation

The variance and standard deviation in the Descriptive Statistics output are both calculated using formulas for unbiased population estimates. The denominator of the variance estimate formula is $n-1$. If you, instead, want the population variance calculated using the formula, $\sigma^2 = \dfrac{\sum(X - \mu)^2}{N}$, it is very simple to use an Excel function to obtain this value.

1. Start with the output obtained in the previous section. Click in cell **E7** and change the label to **Population Standard Dev**. Then click in cell **F7** where the population standard deviation will be placed.

	A	B	C	D	E	F
1	TV	Microwave	Residence		TV	
2	3		1 Apt			
3	0		1 House		Mean	2.636364
4	9		2 Apt		Standard Error	0.777791
5	2		1 House		Median	2
6	3		3 House		Mode	3
7	1		0 House		Population Standard Dev	2.579641

2. Click the **Formulas** tab near the top of the screen and select **Insert Function**.

3. In the Insert Function dialog box, select the **Statistical** category and select the **STDEV.P** function as shown below. Click **OK**.

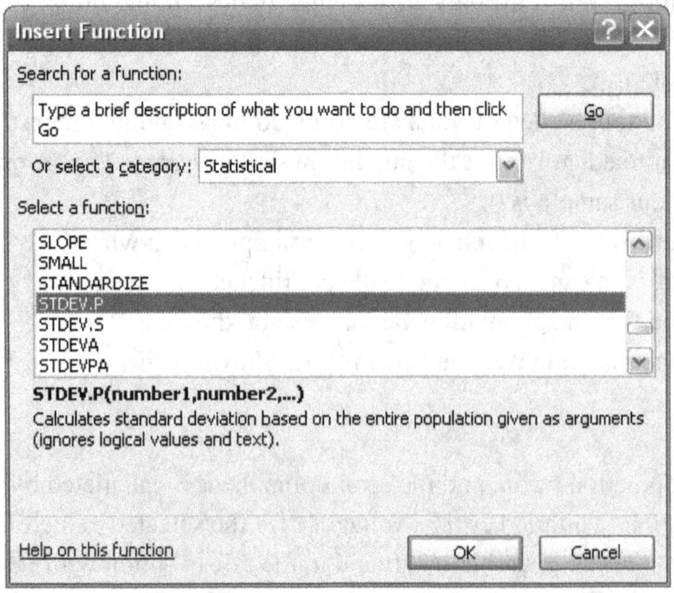

4. Click in the **Number 1** window. Then, in the worksheet, click and drag over the range **A1:A12** to enter the TV data range as shown below. Click **OK**.

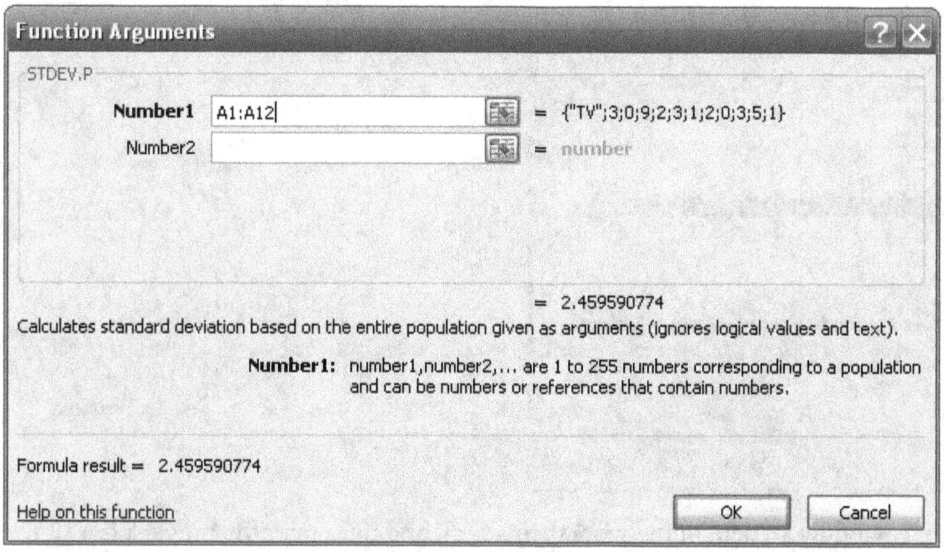

Function Arguments	? X

STDEV.P

Number1	A1:A12	[icon]	= {"TV";3;0;9;2;3;1;2;0;3;5;1}
Number2		[icon]	= number

= 2.459590774

Calculates standard deviation based on the entire population given as arguments (ignores logical values and text).

Number1: number1,number2,... are 1 to 255 numbers corresponding to a population and can be numbers or references that contain numbers.

Formula result = 2.459590774

Help on this function   OK   Cancel

5. The function returns a population standard deviation of 2.4596. Click in cell **E8** and change the label to **Population Variance**. Then click in **F8** where the population variance will be placed.

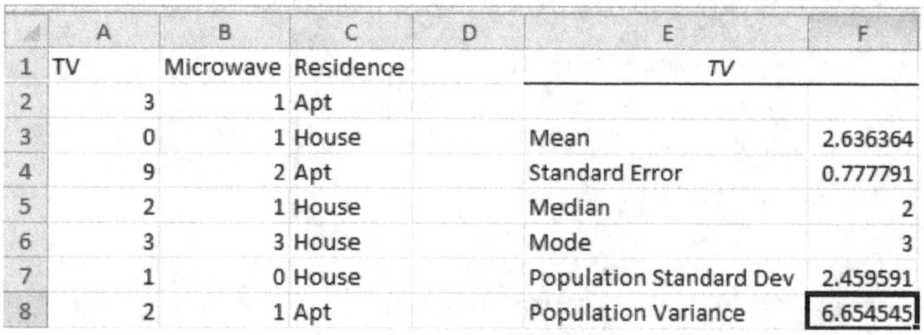

	A	B	C	D	E	F
1	TV	Microwave	Residence		TV	
2	3		1 Apt			
3	0		1 House		Mean	2.636364
4	9		2 Apt		Standard Error	0.777791
5	2		1 House		Median	2
6	3		3 House		Mode	3
7	1		0 House		Population Standard Dev	2.459591
8	2		1 Apt		Population Variance	6.654545

6. Select **Insert Function** located in the Formulas ribbon.

7. In the Insert Function dialog box, select the **Statistical** category, select the **VAR.P** function, and click **OK**.

Insert Function

Search for a function:

| Type a brief description of what you want to do and then click Go | | Go |

Or select a category: Statistical

Select a function:

```
T.TEST
TREND
TRIMMEAN
VAR.P
VAR.S
VARA
VARPA
```

VAR.P(number1,number2,...)
Calculates variance based on the entire population (ignores logical values and text in the population).

Help on this function   OK   Cancel

8. Click in the **Number 1** window. Then, in the worksheet, click and drag over the range **A1:A12** to enter the TV data range. Click **OK**.

Function Arguments

VAR.P

| **Number1** | A2:A12 | | = {3;0;9;2;3;1;2;0;3;5;1} |
| Number2 | | | = number |

= 6.049586777

Calculates variance based on the entire population (ignores logical values and text in the population).

Number1: number1,number2,... are 1 to 255 numeric arguments corresponding to a population.

Formula result = 6.049586777

Help on this function   OK   Cancel

9. The function returns a population variance of 6.0496. Click in cell **F4** to change the value of the standard error.

	A	B	C	D	E	F
1	TV	Microwave	Residence		TV	
2	3	1	Apt			
3	0	1	House		Mean	2.636364
4	9	2	Apt		Standard Error	0.777791

10. When the population standard deviation is known, the standard error is calculated using the formula $\sigma_{\bar{x}} = \dfrac{\sigma}{\sqrt{N}}$. We will key in a formula to obtain this result. In cell F4, key in **=F7/sqrt(F15)**. Press **[Enter]**.

	A	B	C	D	E	F
1	TV	Microwave	Residence		TV	
2	3	1	Apt			
3	0	1	House		Mean	2.636364
4	9	2	Apt			=F7/sqrt(F15)

11. The standard error is equal to 0.7416. Click in cell **F18** where the 95% confidence level will be placed.

12. The 95% confidence level is equal to the product of the standard error times the critical z for a two-tailed test. For this problem, the standard error is found in cell F4 and the critical z is 1.96. We will use a formula to carry out the calculations. In cell F18, key in **=F4*1.96** as shown below. Press [**Enter**].

					E	F
16					Largest(2)	5
17					Smallest(2)	0
18						=F4*1.96

The 95% confidence level is equal to 1.4535. When you want population values instead of population estimates based on a sample, you need to change the population standard deviation, the population variance, the standard error, and the confidence level. The completed worksheet with these population values is shown below.

	A	B	C	D	E	F
1	TV	Microwave	Residence		TV	
2	3	1	Apt			
3	0	1	House		Mean	2.636364
4	9	2	Apt		Standard Error	0.741595
5	2	1	House		Median	2
6	3	3	House		Mode	3
7	1	0	House		Population Standard Dev	2.459591
8	2	1	Apt		Population Variance	6.049587
9	0	2	Apt		Kurtosis	3.212809
10	3	1	House		Skewness	1.602855
11	5	1	House		Range	9
12	1	1	Apt		Minimum	0
13					Maximum	9
14					Sum	29
15					Count	11
16					Largest(2)	5
17					Smallest(2)	0
18					Confidence Level(95.0%)	1.453525

Steps to Follow to Obtain Descriptive Statistics for Two or More Variables

In order to use the Descriptive Statistics Analysis Tool to produce descriptive statistics for two or more variables simultaneously, the variables must be located in either adjacent columns (or adjacent rows) in the worksheet. In this section, I provide a brief explanation of how to obtain descriptive statistics for two variables. If a more complete description is required, you might want to refer back to the steps presented in the first section of this chapter.

1. If you have not already done so, enter the data in an Excel worksheet as shown on the first page of this chapter or open the "Ch5_Energy Use" worksheet on the Web site.

2. Click the **Data** tab near the top of the screen and select **Data Analysis** in the Analysis group.

If Data Analysis does not appear as a choice in the Data ribbon, you will need to load the Microsoft Excel ToolPak add-in. Follow the procedure on page 9.

3. In the Data Analysis dialog box, select **Descriptive Statistics** and click **OK**.

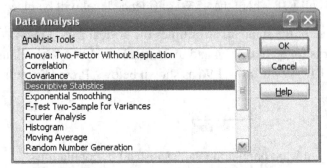

4. Complete the Descriptive Statistics dialog box as shown below. The entries are nearly the same as in the previous section except that the input range includes the worksheet location of both TV and MICROWAVE, **$A$1:$B$12**, the output will be placed in a new worksheet, and we have requested the first largest and first smallest values. Click **OK**.

The output is shown at the top of the next page. Adjust the column width as necessary to display the labels. For information regarding the interpretation of the output, see **Interpreting the Output** on pages 80-82.

	A	B	C	D
1	*TV*		*Microwave*	
2				
3	Mean	2.636364	Mean	1.272727
4	Standard Error	0.777791	Standard Error	0.237062
5	Median	2	Median	1
6	Mode	3	Mode	1
7	Standard Deviation	2.579641	Standard Deviation	0.786245
8	Sample Variance	6.654545	Sample Variance	0.618182
9	Kurtosis	3.212809	Kurtosis	1.649366
10	Skewness	1.602855	Skewness	0.935197
11	Range	9	Range	3
12	Minimum	0	Minimum	0
13	Maximum	9	Maximum	3
14	Sum	29	Sum	14
15	Count	11	Count	11
16	Largest(1)	9	Largest(1)	3
17	Smallest(1)	0	Smallest(1)	0
18	Confidence Level(95.0%)	1.733026	Confidence Level(95.0%)	0.528207

Missing Values

Missing values should be indicated by blank cells. The count will reflect the number of nonblank cells in the range for each variable included in the input range.

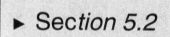

► Section 5.2   Functions: Descriptive Statistics

We can use Excel's functions to obtain the same output that was produced by the Descriptive Statistics Data Analysis Tool, but it requires a little more work. The functions related to descriptive statistics are listed below.

Function Name	Value Returned by the Function
AVEDEV	Average deviation
AVERAGE	Mean
COUNT	Number of observations in a specified range
KURT	Kurtosis
LARGE	Kth observation in a range if values are arranged in descending order
MAX	Maximum
MEDIAN	Median
MIN	Minimum
MODE.MULT	Modes if the range of values has more than one mode. The range must be entered as an array. The result will be returned as an array.
MODE.SNGL	Mode if the range of values has only one mode. The range must be entered as an array.

QUARTILE.INC	First, second, third, and fourth quartile
SKEW	Skewness
SMALL	Kth observation in a range if observations are arranged in ascending order
STDEV.P	Standard deviation of a population calculated using the population formula
STDEV.S	Standard deviation of a sample calculated using the unbiased estimate formula
VAR.P	Variance of a population calculated using the population formula
VAR.S	Variance of a sample calculated using the unbiased estimate formula

Sample Research Problem

To illustrate the use of Excel's functions for obtaining descriptive statistics, I will use the same set of data that was utilized at the beginning of this chapter to illustrate the Descriptive Statistics Analysis Tool.

Steps to Follow to Obtain Descriptive Statistics for One Variable

I have chosen the TV variable to illustrate how to use functions to obtain descriptive statistics for one variable. I will provide detailed instructions for only a couple of the statistics because the procedure is quite repetitive. Once you have the desired descriptive statistics for one variable, obtaining them for additional variables is a relatively simple process, as long as the variables are located in adjacent columns of the worksheet.

1. If you have not already done so, enter the data shown on the first page of this chapter or open the "Ch5_Energy Use" worksheet on the Web site.

2. Let's begin with the average deviation. The average deviation is found by first calculating the absolute value of the deviation score for each observation $|X - \overline{X}|$ and then calculating the average. First let's provide a label for the output. Click in **A14** and key in **AVEDEV**. Then click in cell **B14** where the average deviation will be placed.

Labels are not necessary, but they help to make the output more informative, especially if you refer to it a few days after creating it.

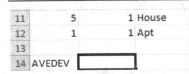

3. Click the **Formulas** tab near the top of the screen and select **Insert Function**.

4. In the Insert Function dialog box, select the **Statistical** category and the **AVEDEV** function. Click **OK**.

Insert Function

Search for a function:

```
Type a brief description of what you want to do and then click
Go
```
[Go]

Or select a category: Statistical

Select a function:

```
AVEDEV
AVERAGE
AVERAGEA
AVERAGEIF
AVERAGEIFS
BETA.DIST
BETA.INV
```

AVEDEV(number1,number2,...)
Returns the average of the absolute deviations of data points from their mean.
Arguments can be numbers or names, arrays, or references that contain
numbers.

Help on this function   [OK]   [Cancel]

5. Click in the **Number 1** window. Then click in cell **A2** of the worksheet where the TV data begin and drag to cell **A12** where the data end. If you prefer, you can manually enter **A2:A12** in the Number 1 window.

Function Arguments

AVEDEV

Number1	A2:A12	= {3;0;9;2;3;1;2;0;3;5;1}
Number2		= number

= 1.785123967

Returns the average of the absolute deviations of data points from their mean. Arguments can be numbers or names, arrays, or references that contain numbers.

Number1: number1,number2,... are 1 to 255 arguments for which you want the average of the absolute deviations.

Formula result = 1.785123967

Help on this function   [OK]   [Cancel]

6. Click **OK**. The AVEDEV function returns a value of 1.7851 and places it in cell B14 of the worksheet.

11	5	1	House
12	1	1	Apt
13			
14	AVEDEV	1.78512397	

7. Next let's find the average (or mean) by using the AVERAGE function. Click in cell **A15** and enter the label **AVERAGE**. Then click in cell **B15** where the average will be placed.

11	5	1	House
12	1	1	Apt
13			
14	AVEDEV	1.78512397	
15	AVERAGE		

8. Click the **Formulas** tab near the top of the screen and select **Insert Function**.

9. In the Insert Function dialog box, select the **Statistical** category and the **AVERAGE** function as shown below. Click **OK**.

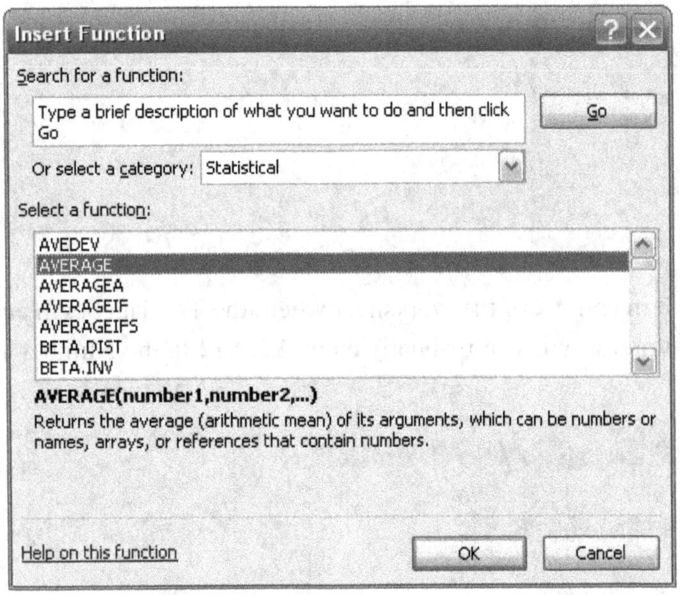

Insert Function   ? X

Search for a function:

Type a brief description of what you want to do and then click Go   Go

Or select a category: Statistical

Select a function:

AVEDEV
AVERAGE
AVERAGEA
AVERAGEIF
AVERAGEIFS
BETA.DIST
BETA.INV

AVERAGE(number1,number2,...)
Returns the average (arithmetic mean) of its arguments, which can be numbers or names, arrays, or references that contain numbers.

Help on this function   OK   Cancel

10. Click in the **Number 1** window. Then click on cell **A2** where the TV data begin and drag to cell **A12** where the TV data end.   If you prefer, you can manually enter **A2:A12** in the Number 1 window.

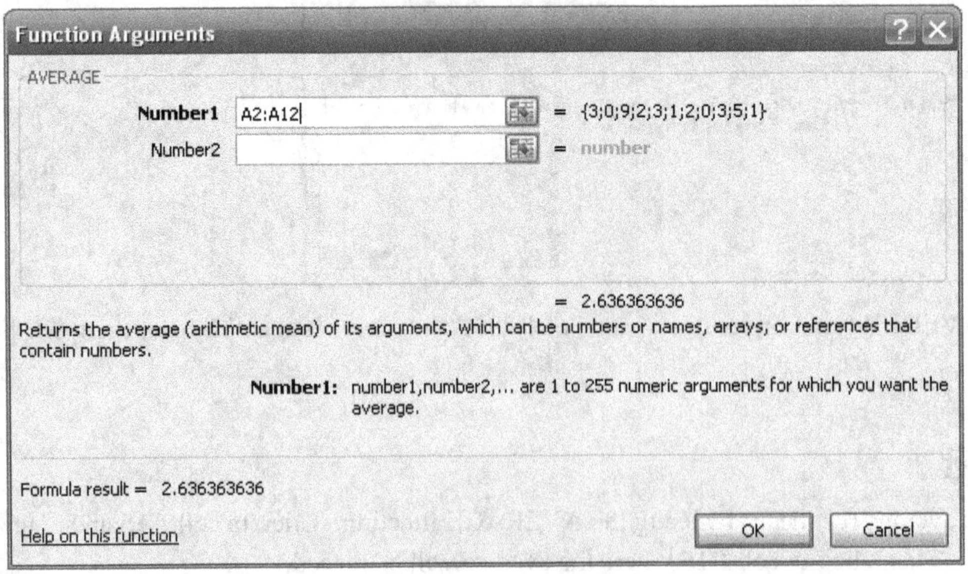

Function Arguments   ? X

AVERAGE

Number1   A2:A12   = {3;0;9;2;3;1;2;0;3;5;1}

Number2   = number

= 2.636363636

Returns the average (arithmetic mean) of its arguments, which can be numbers or names, arrays, or references that contain numbers.

Number1: number1,number2,... are 1 to 255 numeric arguments for which you want the average.

Formula result = 2.636363636

Help on this function   OK   Cancel

11. Click **OK** and a value of 2.6364 is returned and placed in cell B14 of the worksheet.

11	5	1 House
12	1	1 Apt
13		
14	AVEDEV	1.78512397
15	AVERAGE	2.63636364

As you can see, regardless of the function you are using, the steps that you follow are very similar. If you run into any difficulty, you can read the description provided at the bottom of the Insert Function dialog box and at the bottom of the dialog box for a specific function. If you need additional information, just click the question mark in the top right corner of the function's dialog box.

Steps to Follow to Obtain Descriptive Statistics for Two or More Variables

As I said earlier, once you have descriptive statistics for one variable, obtaining them for additional variables is a very simple process, as long as the variables appear in adjacent columns (or rows) in the worksheet. You essentially just copy the functions and formulas from one column in the output table to another. In the previous section, we found the average deviation and the average for TV and placed the values in column B. Now let's obtain those two values for Microwave.

1. When we completed the final step in the last section, the first two columns of our worksheet looked like those shown below.

	A	B	C
1	TV	Microwave	Residence
2	3	1	Apt
3	0	1	House
4	9	2	Apt
5	2	1	House
6	3	3	House
7	1	0	House
8	2	1	Apt
9	0	2	Apt
10	3	1	House
11	5	1	House
12	1	1	Apt
13			
14	AVEDEV	1.78512397	
15	AVERAGE	2.63636364	

Let's continue by finding the average deviation for MICROWAVE and placing it in C14. Click in cell **B14** where the average deviation for TV is located. If you look up in the formula bar, you will see the contents of cell B14. Those contents are =AVEDEV(A2:A12). AVEDEV is the function name and A2:A12 is the data range. If you copy the function in B14 to C14, the column referent will change from A to B, but everything else will stay the same. That's because the cell addresses are relative rather than absolute. So, copy the AVEDEV function in B14 to C14. You should now see a value of 0.5785 in cell C14. Click on C14 to check its contents. In the formula bar you will see =AVEDEV(B2:B12).

14	AVEDEV	1.78512397	0.5785124
15	AVERAGE	2.63636364	

2. Next, let's find the average for MICROWAVE. Click in cell **B15** where the average for TV is located. Look in the formula bar and you will see =AVERAGE(A2:A12). Copy the function in B15 to C15. You should now see a value of 1.2727 in cell C15. Click on C15 to check its contents. In the formula bar you will see =AVERAGE(B2:B12).

| 14 | AVEDEV | 1.78512397 | 0.5785124 |
| 15 | AVERAGE | 2.63636364 | 1.2727273 |

▶ Section 5.3 | Pivot Table: Descriptive Statistics

The Pivot Table is a tool that is very useful for summarizing the variables contained in a database. It is located in the Insert ribbon. Be sure to provide a variable label at the top of each column in the worksheet, because labels are necessary when constructing a Pivot Table.

Sample Research Problem

This research problem is the same one that is presented at the beginning of this chapter. Residential customers of an electric utility were asked to provide information on residence type and the number of TVs and microwaves that they owned.

Steps to Follow to Set up a Pivot Table

Using the Pivot Table, we will be able to obtain the following values: sum, count, average, maximum, minimum, sample standard deviation, population standard deviation, sample variance, and population variance. Let's say that a marketing researcher wanted summary information regarding microwaves for the entire data set and for each residence type separately.

1. If you have not already done so, enter the data shown on the first page of this chapter or open the "Ch5_Energy Use" worksheet on the Web site.

2. Activate any cell in the data range (A1 to C12).

If a cell in the data range is active before Pivot Table is selected, then Excel will automatically enter the data range in the dialog box. You can activate a cell just by clicking in it.

3. Click the **Insert** tab near the top of the screen and select **Pivot Table**. You are given the option of constructing a Pivot Table or a Pivot Chart. Select **Pivot Table**.

4. The range will have been automatically entered into the Table/Range window, and you will see 'Energy Usage'!$A$1:$C$12. The default option for placement of the Pivot Table report output is New Worksheet. Click **OK**.

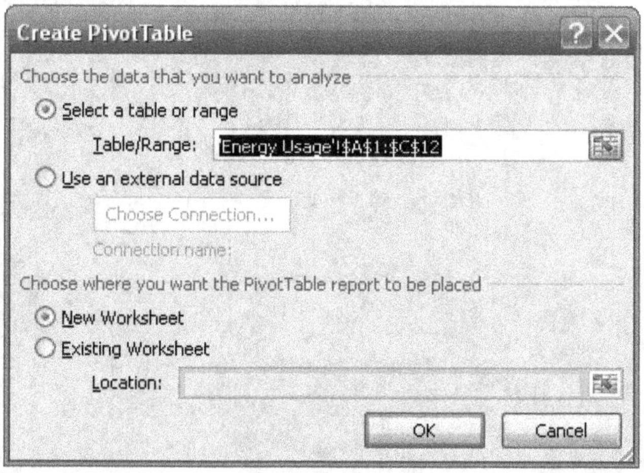

5. On the right side of the worksheet, you will see a Pivot Table Field List. Click the down arrow located above Report Filter and select **Field Section and Areas Section Side-By-Side**.

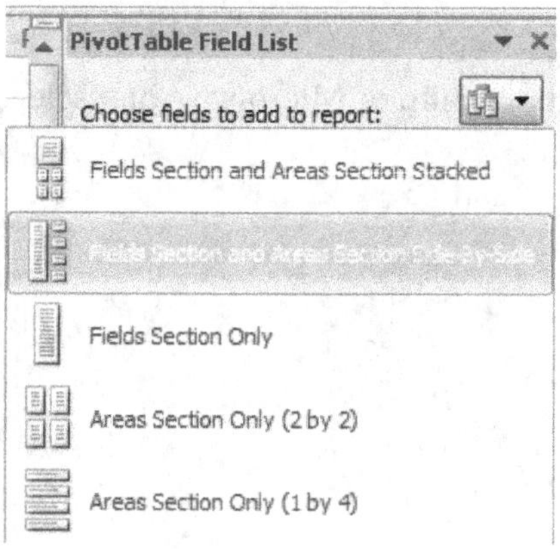

6. Click on **Residence** and drag it to the **Row Labels** area. Click on **Microwave** and drag it to the Σ **Values** area.

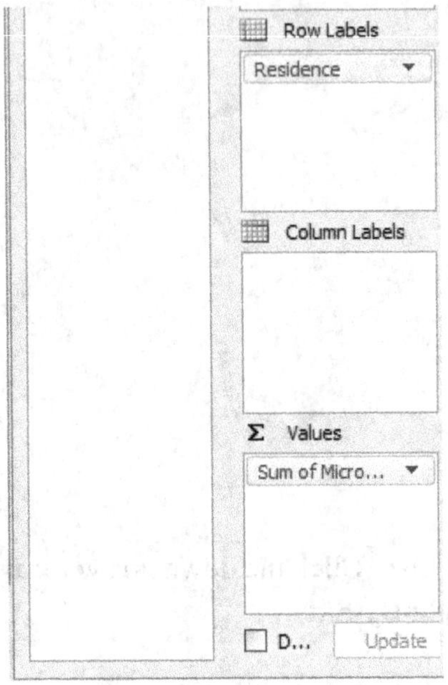

7. First, we will obtain the average. Click the down arrow to the right of Sum of Microwave and select **Value Field Settings** from the shortcut menu.

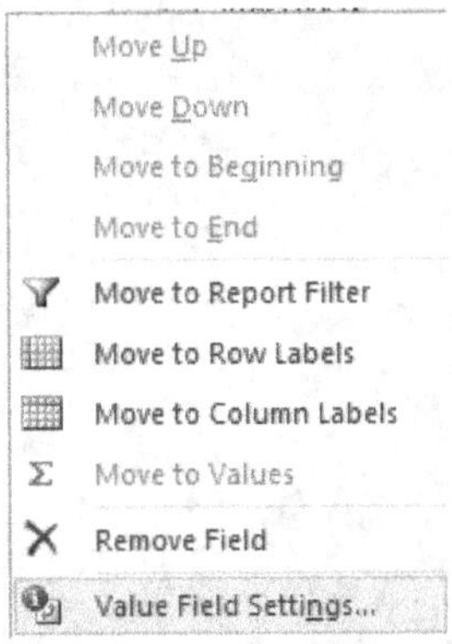

8. Select **Average** as shown at the top of the next page and click **OK**.

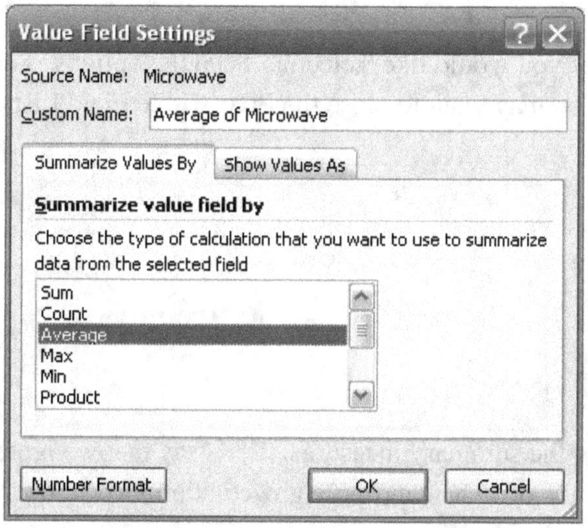

9. Next, let's display the standard deviation. Click on **Microwave** and drag it down to the Σ **Values** area a second time.

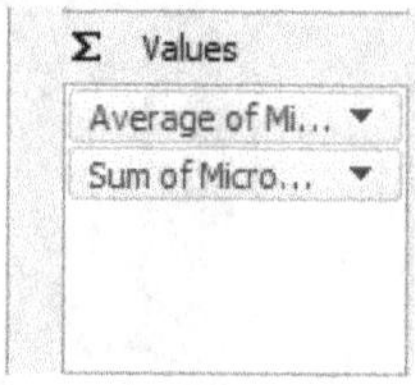

10. Click the down arrow to the right of Sum of Microwave and select **Value Field Settings** from the shortcut menu.

11. Select **StdDev** and click **OK**.

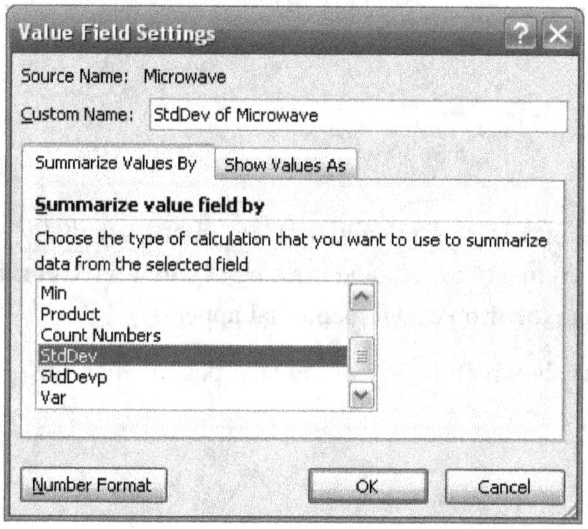

12. You can continue adding the summary measures that you would like, selecting from the options displayed in the Value Field Settings dialog box. You may want to close the Pivot Table Field List on the right so that you can view the entire worksheet.

	A	B	C
1			
2			
3	Row Labels ▾	Average of Microwave	StdDev of Microwave
4	Apt	1.4	0.547722558
5	House	1.166666667	0.98319208
6	Grand Total	1.272727273	0.786245393

13. Excel gives you the option of changing the names of the summary measures. Let's say that we prefer to use the expression *mean* rather than *average*. To make this change, first **right-click** on **Average of Microwave** in cell B3. Select **Value Field Settings** from the drop down menu that appears.

14. Find the Custom Name window at the top of the dialog box. Change the entry in the window from Average of Microwave to **Mean of Microwave**. Click **OK**.

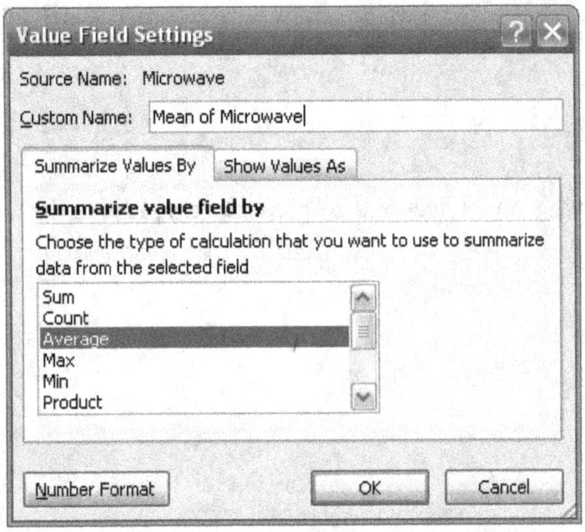

15. The microwave mean is a calculated value that could have several decimal places. Let's say that we want the value displayed with two decimal places. To make this change, first **right-click** on **Mean of Microwave** in cell B3. Select **Number Format** from the drop down menu that appears.

16. Select the **Number** category and **2** decimal places as shown at the top of the next page. Click **OK**.

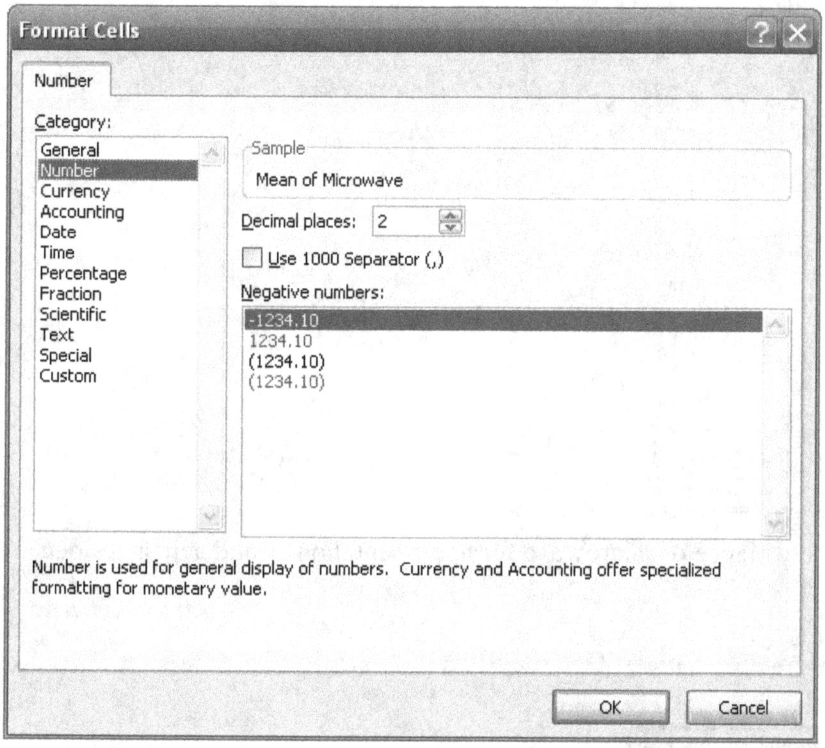

Format Cells

Number

Category:

General
Number
Currency
Accounting
Date
Time
Percentage
Fraction
Scientific
Text
Special
Custom

Sample
Mean of Microwave

Decimal places: 2

☐ Use 1000 Separator (,)

Negative numbers:

-1234.10
1234.10
(1234.10)
(1234.10)

Number is used for general display of numbers. Currency and Accounting offer specialized formatting for monetary value.

OK   Cancel

Your worksheet should look similar to the one shown below, with all means displayed with 2 decimal places.

	A	B	C
1			
2			
3	Row Labels ▾	Mean of Microwave	StdDev of Microwave
4	Apt	1.40	0.547722558
5	House	1.17	0.98319208
6	Grand Total	1.27	0.786245393

As you can see, Pivot Tables are fairly easy to construct. In addition, they can be customized to meet the user's needs. Several summary measures may be displayed in a single Pivot Table, several formats are available (e.g., number, currency, date), and grand totals are optional.

Changing or Removing Summary Measures from a Pivot Table

When you are working with a Pivot Table, you might decide to change the summary measures that are included.

1. Let's say that you decide to replace the standard deviation with the variance in the Pivot Table created in the previous section. **Right-click** on **StdDev of Microwave** in cell C3 and select **Summarize Values By**. Click **More options**. Select **Var** as shown at the top of the next page. Click **OK**.

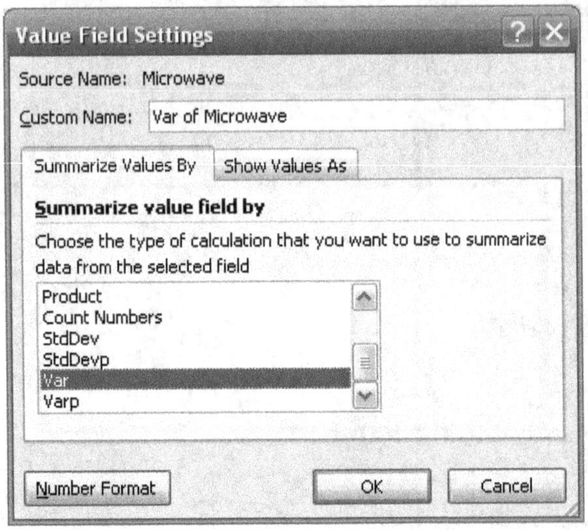

Value Field Settings
Source Name: Microwave
Custom Name: Var of Microwave
Summarize Values By
Show Values As
Summarize value field by
Choose the type of calculation that you want to use to summarize data from the selected field
Product
Count Numbers
StdDev
StdDevp
Var
Varp
Number Format
OK
Cancel

The Pivot Table now displays the variance of microwave for apartment, house, and across residence types.

	A	B	C
1			
2			
3	Row Labels	Mean of Microwave	Var of Microwave
4	Apt	1.40	0.3
5	House	1.17	0.966666667
6	**Grand Total**	**1.27**	**0.618181818**

2. It's also very easy to remove a summary measure. Let's say that you decide that you don't want to include the variance. **Right-click** on **Var of Microwave** in cell C3.

3. Select **Remove "Var of Microwave"** from the menu.

Copy
Format Cells...
Number Format...
Refresh
Sort
Remove "Var of Microwave"
Remove Values

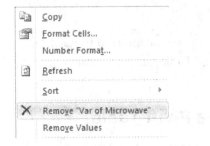

After the variances are removed, your Pivot Table should look like the one shown below.

	A	B
1		
2		
3	Row Labels	Mean of Microwave
4	Apt	1.40
5	House	1.17
6	**Grand Total**	**1.27**

▶ Section 5.4	*Weighting to Adjust for Survey Nonresponse*

Survey researchers often randomly sample potential participants from the population of interest in such a manner that the resulting sample will be representative of the population with respect to important characteristics. For example, researchers may be concerned about the representativeness of their sample with respect to gender, ethnicity, socioeconomic status, or political party affiliation. Even when sampling is done very carefully, the sample may differ substantially from the population on a particular characteristic because members of a subgroup declined to participate. When a researcher is aware that a subgroup is under represented, a decision might be made to weight responses to adjust for survey nonresponse. The example that is provided in this section illustrates how to weight responses for one demographic variable.

> *For more information on weighting survey responses, I recommend you read this 2008 article by Gary R. Pike: Using weighting adjustments to compensate for survey nonresponse, Research in Higher Education, 49, 153-171.*

Sample Research Problem

A university researcher administered a satisfaction survey to graduating seniors to gather data about their satisfaction as a student at the institution. Based on past surveys, the researcher knew that females were more likely to respond and were more satisfied than males. Although the current class of graduating seniors was comprised of 50% females and 50% males, the survey sample was comprised of 65.2% females and 34.8% males. The researcher decided to weight the responses to a survey question that asked about overall satisfaction. Responses were coded as 1 if the student was satisfied and as 0 if the student was not satisfied. For our example, we will work with a data set, shown at the top of the next page, that contains responses from 15 females and 8 males. The average female satisfaction is 12/15 or .8, and the average male satisfaction is 4/8 or .5. Across genders the average satisfaction is 16/23 or .696, a result that puts more emphasis on the females' opinion than on the males'.

Steps to Follow to Weight Responses Based on One Variable

1. Open the "Ch5_Weighting Responses" worksheet on the Web site, or enter the student data shown at the top of the next page in an Excel worksheet.

	A	B	C	D
1	Gender	Response	Weight	Response * Weight
2	Female	1		
3	Female	1		
4	Female	0		
5	Female	1		
6	Female	1		
7	Female	1		
8	Female	1		
9	Female	1		
10	Female	1		
11	Female	0		
12	Female	1		
13	Female	1		
14	Female	0		
15	Female	1		
16	Female	1		
17	Male	1		
18	Male	0		
19	Male	0		
20	Male	1		
21	Male	0		
22	Male	1		
23	Male	0		
24	Male	1		
25	Total	16		

2. You want to weight the responses according to the proportion of females and males in the population.
    To calculate a weight, you will divide the population proportion for each gender subgroup (.5 for both
    females and males) by the subgroup sample size (15 for females and 8 for males) and then multiply this
    result by total sample size (23). If you are using the worksheet available on the Website, you can click
    the sheet tab **Weighting Formulas** at the bottom of the screen and see a worksheet that already has all
    the formulas entered. The worksheet named Values displays the calculated values. If you are not
    working with the Website worksheet, enter the weights in column C as shown at the top of the next
    page.

---

*To display formulas rather than values, click the **Formulas** tab at the top of the screen and select **Show
Formulas** in the Formula Auditing group.*

Trace Precedents   Show Formulas
Trace Dependents   Error Checking
Remove Arrows   Evaluate Formula
        Formula Auditing

	A	B	C
1	Gender	Response	Weight
2	Female	1	=0.5/15*23
3	Female	1	=0.5/15*23
4	Female	0	=0.5/15*23
5	Female	1	=0.5/15*23
6	Female	1	=0.5/15*23
7	Female	1	=0.5/15*23
8	Female	1	=0.5/15*23
9	Female	1	=0.5/15*23
10	Female	1	=0.5/15*23
11	Female	0	=0.5/15*23
12	Female	1	=0.5/15*23
13	Female	1	=0.5/15*23
14	Female	0	=0.5/15*23
15	Female	1	=0.5/15*23
16	Female	1	=0.5/15*23
17	Male	1	=0.5/8*23
18	Male	0	=0.5/8*23
19	Male	0	=0.5/8*23
20	Male	1	=0.5/8*23
21	Male	0	=0.5/8*23
22	Male	1	=0.5/8*23
23	Male	0	=0.5/8*23
24	Male	1	=0.5/8*23

3. The weight applied to the females' responses is 0.7667 and the weight applied to the males' responses is 1.4375. In column D, each survey participant's response will be multiplied by the appropriate weight. Click in cell **D2** and type **B2*C2**. Press [**Enter**].

	A	B	C	D
1	Gender	Response	Weight	Response * Weight
2	Female	1	0.766667	=B2*C2

4. Copy the contents of cell D2 to **D3:D24**.

	A	B	C	D
1	Gender	Response	Weight	Response * Weight
2	Female	1	0.766667	0.766666667
3	Female	1	0.766667	0.766666667
4	Female	0	0.766667	0
5	Female	1	0.766667	0.766666667
6	Female	1	0.766667	0.766666667
7	Female	1	0.766667	0.766666667
8	Female	1	0.766667	0.766666667
9	Female	1	0.766667	0.766666667
10	Female	1	0.766667	0.766666667
11	Female	0	0.766667	0
12	Female	1	0.766667	0.766666667
13	Female	1	0.766667	0.766666667
14	Female	0	0.766667	0
15	Female	1	0.766667	0.766666667
16	Female	1	0.766667	0.766666667
17	Male	1	1.4375	1.4375
18	Male	0	1.4375	0
19	Male	0	1.4375	0
20	Male	1	1.4375	1.4375
21	Male	0	1.4375	0
22	Male	1	1.4375	1.4375
23	Male	0	1.4375	0
24	Male	1	1.4375	1.4375
25	Total	16		14.95

5. The unweighted sum of the responses in column B is 16, and the unweighted average satisfaction across genders is 16/23 or .696. The weighted sum of the responses in column D is 14.95, and the weighted average across genders is 14.95/23 or .65.

Chapter 6

Probability Distributions

Excel provides functions that can be utilized for calculating the probabilities associated with a number of different distributions. Because the required hand calculations are often quite time consuming, you will appreciate the speed with which Excel can compute the probabilities for you. In this chapter, I first present discrete probability distributions and then continuous probability distributions.

▶ Section 6.1	**Discrete Probability Distributions**

Discrete variables can assume, on any single trial, a countable set of values. A variable is considered a discrete random variable if the probability of all possible values can be specified. Introductory statistics textbooks often use coin tossing and random selection of research participants from a subject pool to illustrate probabilities of discrete random variables. I will describe how Excel's functions can be used to answer questions regarding three discrete probability distributions: binomial, hypergeometric, and Poisson.

Binomial Distribution

Binomial variables take on only two values. One of these values is typically designated as a "success" and the other as a "failure." The probability of a "success" is symbolized as p, and the probability of a "failure" as q. Because "success" and "failure" are the only possible outcomes, the sum of their probabilities is 1.

We will use the example of tossing a fair coin five times to illustrate how to utilize Excel's functions for computing probabilities for a binomial distribution. We will consider "heads" a success and "tails" a failure. Because we are tossing a fair coin, the probability of a success on a single trial is .5. The outcomes are independent, because the probability of tossing heads on one trial does not change the probability of tossing heads on a subsequent trial.

The general form of the probability function for the binomial distribution is given by

$$P(X = x) = \binom{N}{x} p^x q^{N-x}$$

where $P =$ probability.

$X =$ the event of interest. For our example, the event of interest is tossing heads.

$x =$ the specific number of times the event of interest occurs. For our example of five tosses, x can take on values from 0 to 5.

$N =$ the number of tosses (or trials). For our example, $N = 5$ tosses.

$p =$ the probability of success. For our example, success is tossing heads. Because the coin is fair, $p = .5$.

q = the probability of failure. For our example, failure is tossing tails. Because $(p + q) = 1$ and $p = .5$, q must also be equal to .5.

$P(X = x)$, therefore, stands for the probability that the number of tosses that result in heads is equal to a specific number (0, 1, 2, 3, 4, or 5).

We will display the probabilities for this binomial distribution in an Excel worksheet.

Binominal Probabilities

1. Complete an Excel worksheet with entries as shown below or open worksheet "Ch6_Binomial" on the Web site. If you are entering the data manually, be sure to include the labels, **X** and **P(X=x)**, in cells A1 and B1, respectively.

	A	B
1	X	P(X=x)
2	0	
3	1	
4	2	
5	3	
6	4	
7	5	

2. The probability that $X = 0$, $P(X = 0)$, will be placed in cell B2 of the output table, so activate cell **B2**.

3. Click the **Formulas** tab near the top of the screen and select **Insert Function**.

4. In the Insert Function dialog box, select the **Statistical** function category and the **BINOM.DIST** function. Click **OK**.

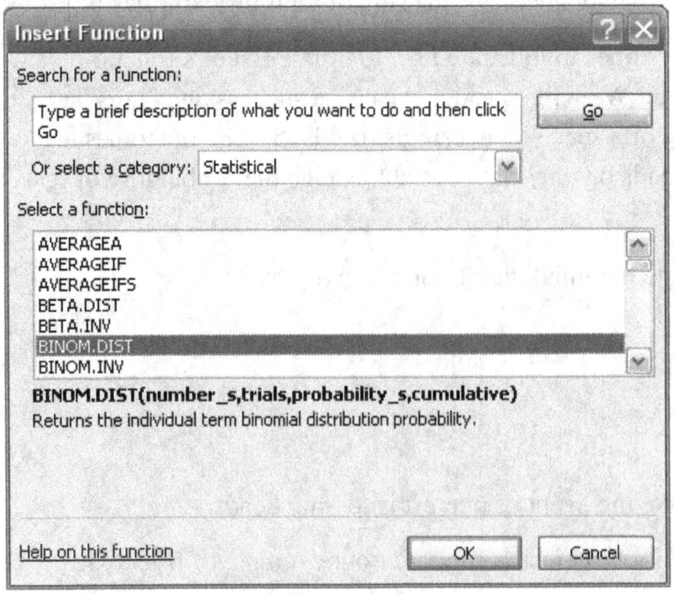

5. Complete the BINOM.DIST dialog box as shown below. Details regarding the entries are given immediately after the dialog box.

- **Number_s**. The number of successes is equal to 0. We could enter 0 here, but I recommend that we enter a cell address, because that will enable us to copy the function to the other cells in the output table. That will save us some work later on. So click in the **Number_s** window and enter **A2**. (A2 is the cell location of 0.)

- **Trials**. The total number of trials (tosses) is equal to 5. Click in the **Trials** window and enter **5**.

- **Probability_s**. The probability of a success, or tossing heads with a fair coin, is .5. Click in the **Probability_s** window and enter **.5**.

- **Cumulative**. We do not want cumulative probabilities in column B of the worksheet, so enter **false**.

6. Click **OK** and a value of .03125 is returned. This is the probability of tossing a fair coin five times and having none of the tosses result in a head (0 successes).

	A	B
1	X	P(X=x)
2	0	0.03125
3	1	
4	2	
5	3	
6	4	

7. To obtain the remaining probabilities, we can copy the **BINOM.DIST** function that was entered in cell B2 to cells B3 through B7. Activate cell **B2**. Look in the Formula Bar to make sure that the cell entry is the BINOM.DIST function, BINOM.DIST (A2, 5, 0.5, false).

8. Use the fill handle to copy the BINOM.DIST function. To do this, move the pointer to the lower right corner of cell B2. It will turn into a black plus sign. Press the left mouse key and hold it down while you drag to cell B7. Release the mouse key and the probabilities for 1 to 5 successes will be returned. Check the accuracy of your output by referring to the worksheet shown at the top of the next page.

	A	B
1	X	P(X=x)
2	0	0.03125
3	1	0.15625
4	2	0.3125
5	3	0.3125
6	4	0.15625
7	5	0.03125

Cumulative Binomial Probabilities

The BINOM.DIST function can also be used to obtain cumulative probabilities. We will use the same output table that we created in the last section.

1.   In cell **C1** of the output table, enter the label **P(X<=x)**. This stands for the probability that X (the event of interest) is less than or equal to x (specific values ranging from 0 to 5).

	A	B	C
1	X	P(X=x)	P(X<=x)

2.   The probability that X is less than or equal to 0, **P(X<=0)**, is a cumulative probability. Activate cell **C2** where this cumulative probability value will be placed.

3.   Select **Insert Function** in the Formulas ribbon at the top of the screen.

4.   In the Insert Function dialog box, select the **Statistical** category and select the **BINOM.DIST** function. Click **OK**.

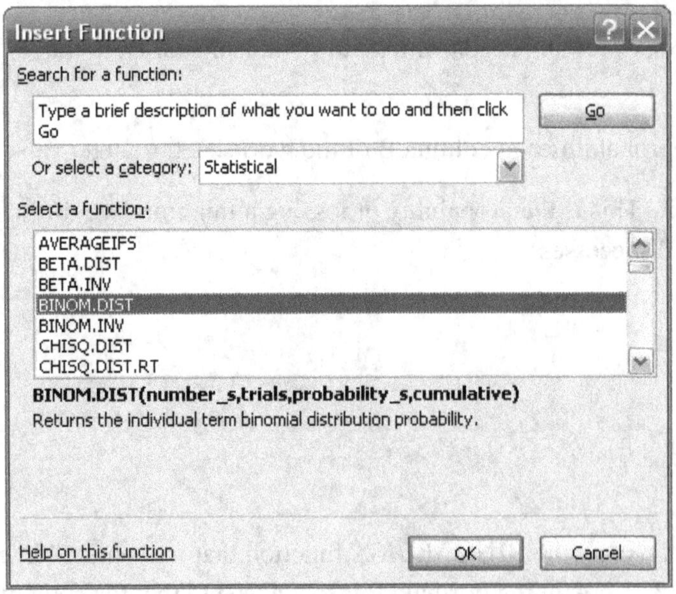

5. Complete the BINOM.DIST dialog box as shown below. Detailed information regarding the entries is given immediately after the dialog box.

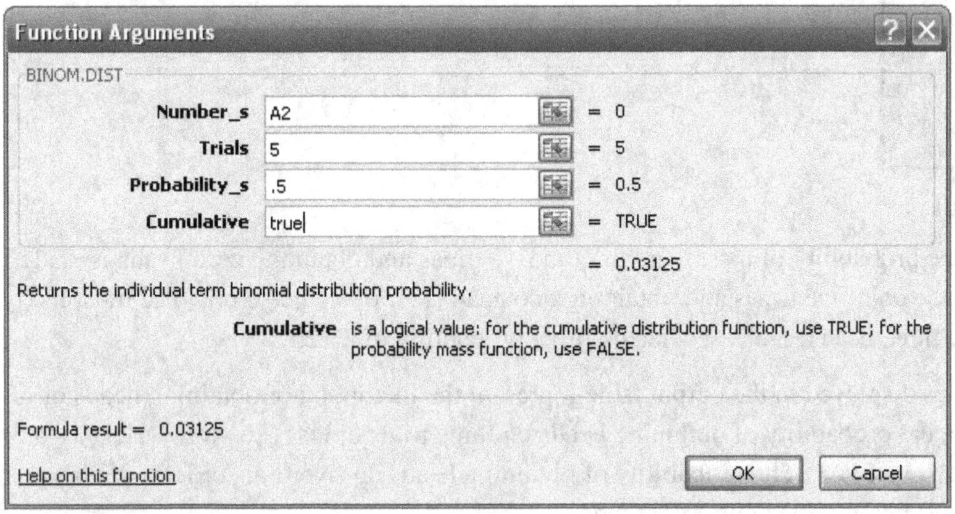

- **Number_s**. Click in the **Number_s** window and enter **A2**, the cell location of 0 (the number of successes).

- **Trials**. Click in the **Trials** window and enter **5**, the total number of trials (tosses).

- **Probability_s**. Click in the **Probability_s** window and enter **.5**, the probability of a success (heads) on a single trial.

- **Cumulative**. Click in the **Cumulative** window and enter **true**, because we want the cumulative probabilities in Column C of the output table.

6. Click **OK** and a value of .03125 is returned.

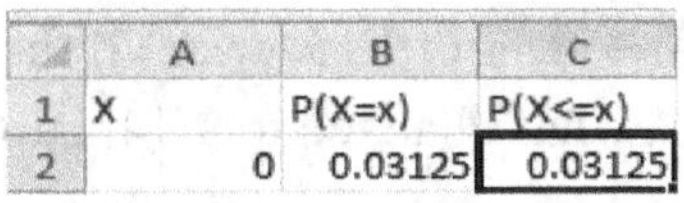

	A	B	C
1	X	P(X=x)	P(X<=x)
2		0	0.03125	0.03125

7. **Cumulative probabilities for 1 to 5 successes**. To obtain the remaining cumulative probabilities, we will copy the BINOM.DIST function in cell C2 to cells C3 through C7. Activate cell **C2** where the cumulative probability associated with zero successes was placed.

Note that it is essential that a cell address (A2) is entered in the dialog box for Number_s in order for us to copy the function accurately.

8. Use the fill handle to copy the BINOM.DIST function. To do this, move the pointer to the lower right corner of cell C2. It will turn into a black plus sign. Press the left mouse key and hold it down while you drag to cell C7. Release the mouse key and the cumulative probabilities for 1 to 5 successes will be returned. Check the accuracy of your output by referring to the worksheet shown at the top of the next page.

	A	B	C
1	X	P(X=x)	P(X<=x)
2	0	0.03125	0.03125
3	1	0.15625	0.1875
4	2	0.3125	0.5
5	3	0.3125	0.8125
6	4	0.15625	0.96875
7	5	0.03125	1

Interpreting the Output

For our sample problem, the probability of tossing a fair coin five times and obtaining zero heads is .03125. The probability of tossing the coin five times and obtaining one head is .15625, and so on. The most likely outcomes are two heads or three heads, each associated with a probability of .3125.

The cumulative probabilities displayed in the output table represent the likelihood of obtaining heads on x trials or less. For example, the probability of obtaining heads on three trials or less (i.e., three trials, two trials, one trial, or zero trials) is .8125. The probability of obtaining heads on five trials or less, of course, is equal to 1 because the coin is being tossed only five times.

Hypergeometric Distribution

The HYPGEOM.DIST function returns probabilities associated with the hypergeometric distribution. The hypergeometric distribution is similar to the binomial distribution except we do not have independence. More specifically, the outcome on one trial changes the probability of an outcome on a subsequent trial. We will use an example of randomly sampling participants for a research project to illustrate how to use the HYPGEOM.DIST function. Let's say that 15 volunteers signed up to participate in the research project but we only need three subjects. We know that nine of the volunteers are women and that six are men. If we randomly sample three volunteers, what is the probability that two will be women and one will be a man? In this situation, we are sampling without replacement, because after a person is selected we do not want to select that person again. Before selection begins, the probability of sampling a woman is 9/15, or .6. If the first draw is a man, then the probability of sampling a woman on the second draw has changed to 9/14, or .64, because only 14 persons remain in the subject pool.

The general form of the probability function for the hypergeometric distribution is given by

$$P(X=x) = \frac{\binom{M}{x}\binom{N-M}{n-x}}{\binom{N}{n}}$$

where M = the number of successes in the population. The number of women in the population is 9.

x = the number of successes in the sample. We will declare selecting a woman a success.

N = the size of the population. For our example, N = 15 volunteers.

n = the size of the sample. We are drawing a sample of size n = 3.

We will display the probabilities for this research participant sampling problem in an Excel worksheet.

1. Complete the worksheet with entries as shown below or open worksheet "Ch6_Hypergeometric" on the Web site. The X in this worksheet refers to selecting a woman as a research project participant. Because we are randomly sampling a total of three subjects, the specific values this event can take on are 0, 1, 2, and 3. P(X = x) refers to the probabilities associated with each of these specific outcomes.

	A	B
1	X	P(X=x)
2	0	
3	1	
4	2	
5	3	

2. The probability that the sample will contain zero women, P(X = 0), will be placed in cell B2 of the table, so activate cell **B2**.

3. Click **Insert Function** in the Formulas ribbon.

4. In the Insert Function dialog box, select the **Statistical** category and the **HYPGEOM.DIST** function. Click **OK**.

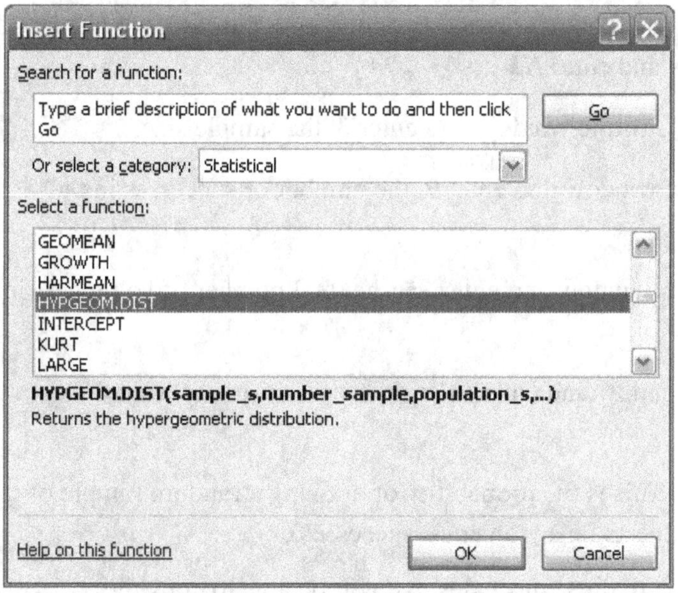

5. Complete the HYPGEOM.DIST dialog box as shown at the top of the next page. Detailed instructions are given immediately following the dialog box.

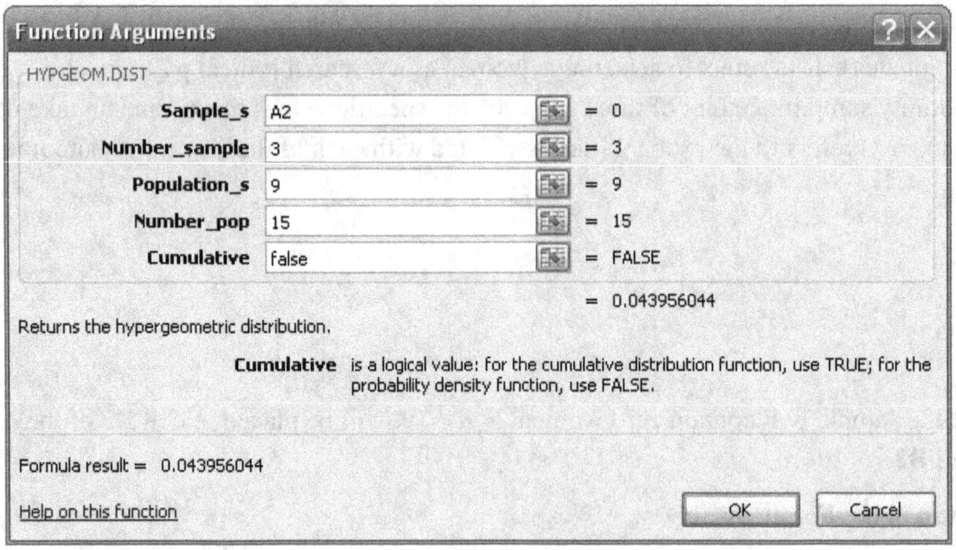

- **Sample_s**. The number of successes (i.e., number of women) in the sample is equal to zero for the problem of finding *P(X = 0)*. Rather than entering 0 in this space, I suggest that you enter the cell address (A2) so that you will be able to copy the function to obtain the probabilities for *x* values 1, 2, and 3. So, click in the **Sample_s** window and enter **A2**.

- **Number_sample**. Click in the **Number_sample** window and enter **3**, the sample size.

- **Population_s**. Click in the **Population_s** window and enter **9**, the number of successes in the population.

- **Number_pop**. Click in the **Number_pop** window and enter **15**, the total number of volunteers in the population.

- **Cumulative**. Click in the **Cumulative** window and enter **false**, because we do not want the cumulative probabilities.

6. Click **OK** and a value of .0440 is returned. This is the probability of drawing a random sample of three research subjects and having none of the subjects be a woman (0 successes).

7. To obtain the remaining probabilities for one to three successes, we will be able to copy the HYPGEOM.DIST function that was entered in cell B2 to cells B3 through B5. Recall that we can copy the function only because we entered a cell address for Sample_s in the HYPGEOM.DIST dialog box when we requested the probability for zero successes. Activate cell **B2**.

8. Copy the function that appears in cell B2 to cells B3 through B5. Check the accuracy of your work by referring to the completed worksheet shown below.

	A	B
1	X	P(X=x)
2	0	0.043956
3	1	0.296703
4	2	0.474725
5	3	0.184615

Interpreting the Output

Given the parameters of our research participant selection problem, the probability that a sample of size $n = 3$ will contain zero women is .0440. The probability that a sample of size $n = 3$ will contain one woman is .2967, and so on. The most likely outcome is a sample with two women and one man, a combination that has a probability of .4747.

Poisson Distribution

The Poisson distribution is especially applicable in situations where the population is very large and the probability associated with a particular event is very small. The general expression for the probability function of the Poisson distribution is given by

$$P(X) = \frac{e^{-\lambda} \lambda^{X}}{X!}$$

where e = a mathematical constant equal approximately to 2.7183,

λ = the expected number of successes, and

X = the number of events.

To illustrate the use of the Poisson distribution, let's assume that the annual rate of mononucleosis among college students on campuses in the United States is 1 in 10,000 students, or .0001. On one particular campus with an enrollment of 15,000, ten students were diagnosed as having mononucleosis. What is the probability of ten such cases if the true rate is .0001? Should the college administrators be alarmed?

Poisson Probabilities

1. Open the worksheet "Ch6_Poisson" on the Web site, or complete a worksheet with entries as shown below. We will use this as an output table to display our work.

	A	B
1	X	P(X)
2	0	
3	1	
4	2	
5	3	
6	4	
7	5	
8	6	
9	7	
10	8	
11	9	
12	10	

2. Let's start with $X = 0$ and then use the copy command to obtain the probabilities associated with X values of 1 through 10. The probability that $X = 0$ will be displayed in cell B2, so activate cell **B2**.

3. Select **Insert Function** in the Formula ribbon.

4. In the Insert Function dialog box, select the **Statistical** category and the **POISSON.DIST** function. Click **OK**.

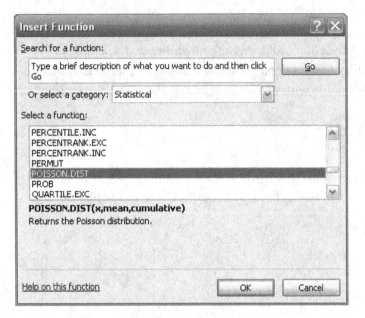

5. Complete the POISSON.DIST dialog box as shown below. Detailed instructions are given immediately following the dialog box.

- **X** is the number of times the event of interest (cases of mononucleosis) occurs. We are computing the probability for *X* = 0 cases. Rather than entering a zero, however, we will enter A2, the cell location of zero in our output table. Click in the **X** window and enter **A2**.

- The **Mean** is the expected value assuming the true rate is .0001. To compute the expected value, we multiply the population size (15,000) times the true rate (.0001), which gives us 1.5. Click in the **Mean** window and enter **1.5**.

- The **Cumulative** window will contain either true or false, depending on whether or not we want cumulative probabilities. We do not want cumulative probabilities. So click in the **Cumulative** window and enter **false**.

6. Click **OK** and a value of .2231 is returned.

	A	B
1	X	P(X)
2	0	0.22313

7. To find the probabilities for 1 to 10 cases of mononucleosis on this particular college campus, we will copy the POISSON.DIST function that was entered in cell B2 of the output table to cells B3 through B12. (Note that *X* had to be specified by a cell address in the function arguments in order for us to copy it and obtain correct values.) Activate cell **B2.**

8. Copy the POISSON.DIST function in cell B2 to cells B3 through B12.   Check the accuracy of your work by referring to the completed worksheet shown below. You will see that the probabilities associated with 9 and 10 cases are displayed in scientific notation. This is because the probabilities are very small.

	A	B
1	X	P(X)
2	0	0.22313
3	1	0.334695
4	2	0.251021
5	3	0.125511
6	4	0.047067
7	5	0.01412
8	6	0.00353
9	7	0.000756
10	8	0.000142
11	9	2.36E-05
12	10	3.55E-06

Cumulative Poisson Probabilities

We can also use Excel's POISSON.DIST function to obtain cumulative probabilities. The cumulative probabilities represent the likelihood that *X* is less than or equal to a specified value. For example, what is the probability that the number of cases of mononucleosis on a campus with 15,000 students will be less than or equal to 5 if the true rate is .0001? We will use the same output table that was created in the previous section.

1. **Label.** In cell **C1** of the output table, enter the label **P(X<=x)**. This stands for the probability that *X* (cases of mononucleosis) is less than or equal to *x* (specific values ranging from 0 to 10).

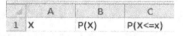

	A	B	C
1	X	P(X)	P(X<=x)

2. Let's begin with P(X <= 0). Activate cell **C2** in the output table.

3. Click **Insert Function** in the Formulas ribbon.

4. In the Insert Function dialog box, select the **Statistical** category and the **POISSON.DIST** function as shown at the top of the next page. Click **OK**.

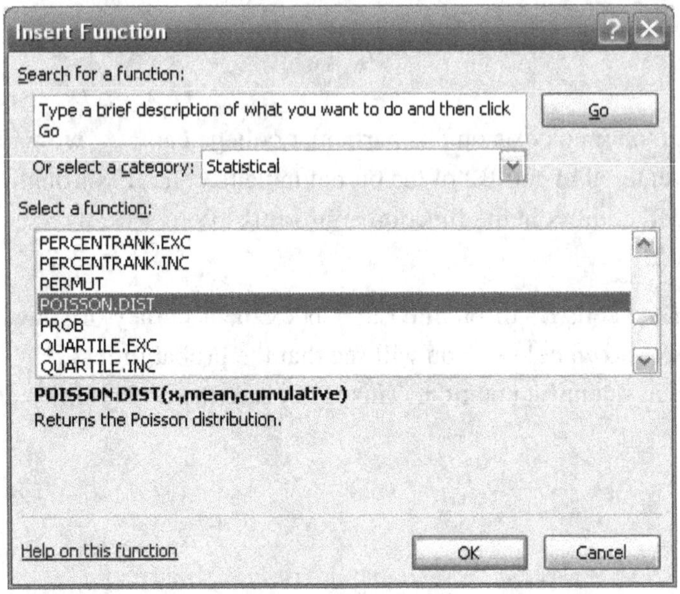

Insert Function   ? X

Search for a function:

```
Type a brief description of what you want to do and then click
Go
```
Go

Or select a category: Statistical

Select a function:

```
PERCENTRANK.EXC
PERCENTRANK.INC
PERMUT
POISSON.DIST
PROB
QUARTILE.EXC
QUARTILE.INC
```

POISSON.DIST(x,mean,cumulative)
Returns the Poisson distribution.

Help on this function   OK   Cancel

5. Complete the POISSON.DIST dialog box as shown below. Detailed information regarding the entries is given immediately following the dialog box.

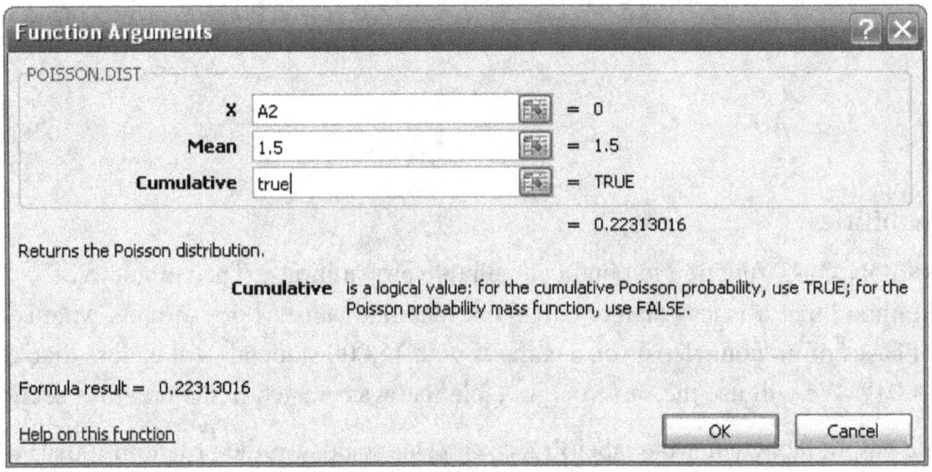

Function Arguments   ? X

POISSON.DIST

X	A2	= 0
Mean	1.5	= 1.5
Cumulative	true	= TRUE

= 0.22313016

Returns the Poisson distribution.

Cumulative is a logical value: for the cumulative Poisson probability, use TRUE; for the Poisson probability mass function, use FALSE.

Formula result = 0.22313016

Help on this function   OK   Cancel

- **X**. Click in the **X** window and enter **A2**, the cell address for 0.

- **Mean**. Click in the **Mean** window and enter the expected value, **1.5**.

- **Cumulative**. Click in the **Cumulative** window and enter **true** to indicate that a cumulative probability is desired.

6. Click **OK** and a value of .2231 is returned.

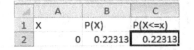

	A	B	C
1	X	P(X)	P(X<=x)
2		0	0.22313	0.22313

7. To obtain the remaining cumulative probabilities, we will copy the POISSON.DIST function in cell C2 to cells C3 through C12. Recall that a cell address must have been entered for X in the POISSON.DIST function dialog box for $X = 0$ in order for the function to copy accurately. Activate cell **C2** where the cumulative probability associated with zero was placed.

8. Copy the cumulative POISSON.DIST function in cell C2 to cells C3 through C12. Check the accuracy by referring to the output table shown below.

	A	B	C
1	X	P(X)	P(X<=x)
2	0	0.22313	0.22313
3	1	0.334695	0.557825
4	2	0.251021	0.808847
5	3	0.125511	0.934358
6	4	0.047067	0.981424
7	5	0.01412	0.995544
8	6	0.00353	0.999074
9	7	0.000756	0.99983
10	8	0.000142	0.999972
11	9	2.36E-05	0.999996
12	10	3.55E-06	0.999999

Interpreting the Output

If the true rate of occurrence of mononucleosis is .0001, then, on a campus of 15,000 students, the probability of zero cases of mononucleosis is .2231. The most likely number of occurrences is 1, which is associated with a probability of .3347. The least likely number of occurrences (of the values we included in the output table) is 10, which is associated with a probability of 3.55E-06, or .00000355. Therefore, the campus administrators should be alarmed if as many as 10 cases are reported.

The probability of one case or less of mononucleosis on this campus is .5578. The probability of two cases or less jumps to .8088.

▶ Section 6.2   # Continuous Probability Distributions

Continuous probability distributions play a major role in statistics because most of the quantitative variables analyzed by researchers are measured on a continuous scale. Furthermore, continuous probability distributions often provide very good approximations for discrete random variables. Unlike discrete variables, continuous variables can take on an infinite number of values between two points. Good examples of such variables are weight, distance, and time. I have included four continuous probability distributions in this section: normal, t, F, and chi-square. I devote most attention to the normal distribution, and only briefly explain how to use Excel's functions in place of textbook tables for the t, F, and chi-square distributions.

Normal Distribution

Excel provides several functions related to the normal distribution.

NORM.DIST returns the normal cumulative distribution for specified X, mean, and standard deviation.

NORM.INV returns the inverse of the normal cumulative distribution for specified probability, mean, and standard deviation.

NORM.S.DIST returns the standard normal cumulative distribution for a specified z value.

NORM.S.INV returns the inverse of the standard normal cumulative distribution for a specified *z* value.

STANDARDIZE returns a standardized score value for specified X, mean, and standard deviation.

In the examples that follow, I will illustrate how to use NORM.S.DIST, NORM.S.INV, and STANDARDIZE.

Calculating Standard Scores Using the STANDARDIZE Function

For our example, let's use the height of adult women in the U.S., measured in inches. Assume that the mean (μ) of the population is 64 inches (i.e., 5'4") and the standard deviation is 3 inches. Let's transform a height of 58 inches to a standardized value.

1.   Open a new, blank Excel worksheet and click in cell **A1**.

2.   Click the **Formulas** tab near the top of the screen and select **Insert Function**.

3.   In the Insert Function dialog box, select the **Statistical** category and the **STANDARDIZE** function as shown below. Click **OK**.

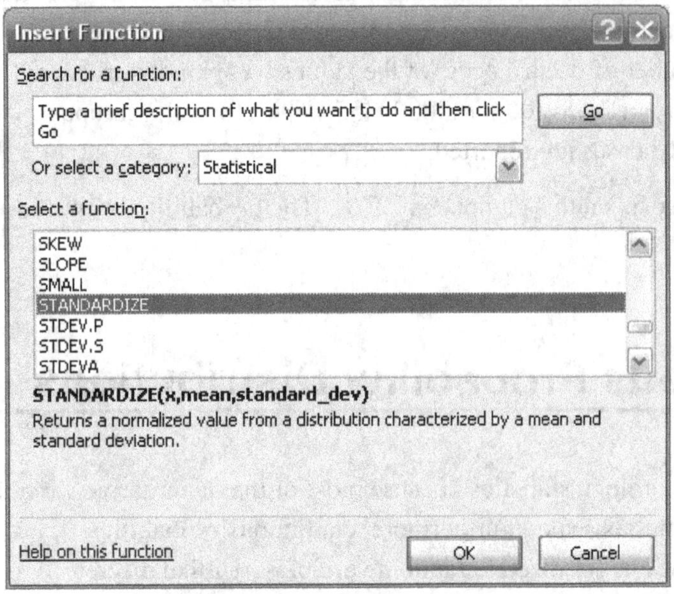

4.   Complete the STANDARDIZE dialog box as shown at the top of the next page. Detailed instructions are given immediately following the dialog box.

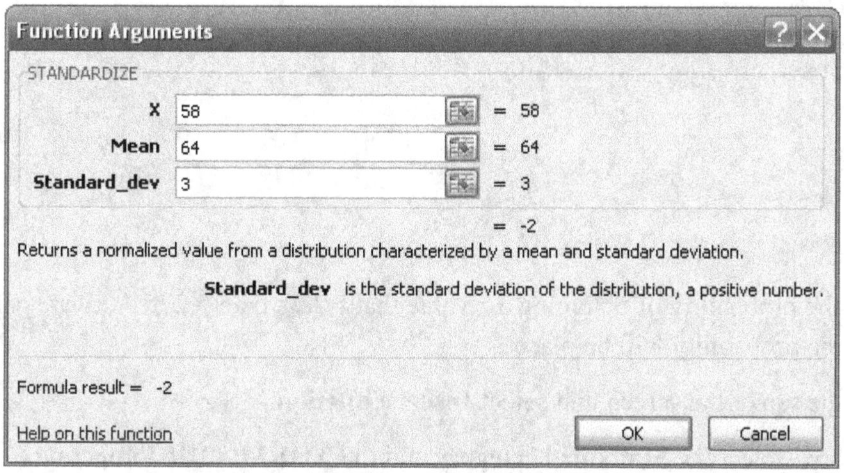

- **X**. Enter **58**, the height you want to transform to a standardized value.

- **Mean**. Enter **64**, the population mean.

- **Standard_dev**. Enter **3**, the population standard deviation.

5. Click **OK** and a z of -2 is returned.

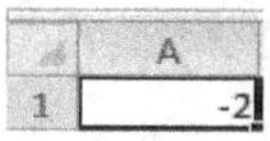

Calculating Cumulative Probabilities Using the NORM.S.DIST Function

If we are working with variables that are distributed normally, one of the advantages of standardized scores is that we can easily determine the probability of obtaining a z value that is less than, or greater than, a specified value. To illustrate the computation of cumulative probabilities, we will be using the NORM.S.DIST function. (If you are working with raw score values, you will use the NORM.DIST function, which requires a couple more entries in the dialog box.)

The NORM.S.DIST works about the same as the normal distribution table found in the back of most statistics textbooks, only better. With textbook tables we are usually limited to z values displayed with a maximum of two decimal places and must interpolate to obtain the probabilities associated with z values with more than two decimals. To illustrate the beauty of Excel for the task of calculating cumulative probabilities, I have included in the example some z values with four decimals, as well as a couple of extreme z values not found in most textbook tables.

1. Open worksheet "Ch6_Normsdist" on the Web site, or prepare an output table by entering the column labels, **Z, P(Z<z), P(Z>z)**, and the specific z values shown in column A at the top of the next page. Make columns B and C wide enough so that several decimal places can be displayed.

	A	B	C
1	Z	P(Z<z)	P(Z>z)
2	5.27		
3	2.6725		
4	1		
5	-1		
6	-2.6725		
7	-5.27		

2. We will begin by calculating the probability of obtaining a z value that is less than 5.27. Activate cell **B2** of the output table where the probability will be placed.

3. Click the **Formulas** tab near the top of the screen and select **Insert Function**.

4. In the Insert Function dialog box, select the **Statistical** category and the **NORM.S.DIST** function. Click **OK**.

Insert Function

Search for a function:

| Type a brief description of what you want to do and then click Go | Go |

Or select a category: Statistical

Select a function:

NEGBINOM.DIST
NORM.DIST
NORM.INV
NORM.S.DIST
NORM.S.INV
PEARSON
PERCENTILE.EXC

NORM.S.DIST(z,cumulative)
Returns the standard normal distribution (has a mean of zero and a standard deviation of one).

Help on this function   OK   Cancel

5. Click in the **Z** window and then click in cell **A2** of the worksheet to enter the worksheet address of 5.27 in the dialog box. Click in the **Cumulative** window and enter **true**.

Function Arguments

NORM.S.DIST

| **Z** | A2 | = 5.27 |
| **Cumulative** | true | = TRUE |

= 0.999999932

Returns the standard normal distribution (has a mean of zero and a standard deviation of one).

Cumulative is a logical value for the function to return: the cumulative distribution function = TRUE; the probability density function = FALSE.

Formula result =   0.999999932

Help on this function   OK   Cancel

6. Click **OK** and .999999932 is returned. This is the probability of obtaining a Z value less than 5.27 in the standard normal cumulative distribution.

	A	B	C
1	Z	P(Z<z)	P(Z>z)
2	5.27	0.999999932	

7. To obtain the cumulative probabilities for the remaining z values, first activate cell **B2** where the cumulative probability associated with 5.27 was placed. Then use the copy command to copy the NORM.S.DIST function in cell B2 to cells B3 through B7. A completed output table is displayed below. As you can see, we can quickly obtain cumulative probabilities for familiar z values such as 1 and -1, as well as for z values with several decimal places (e.g., 2.6725).

	A	B	C
1	Z	P(Z<z)	P(Z>z)
2	5.27	0.999999932	
3	2.6725	0.996235581	
4	1	0.841344746	
5	-1	0.158655254	
6	-2.6725	0.003764419	
7	-5.27	6.82119E-08	

8. The column C probabilities are easy to obtain after completing column B. We will utilize the formula: 1 minus the value in column B. Activate cell **C2** where the first cumulative probability will be placed. Enter the formula **=1-B2** and press [**Enter**].

	A	B	C
1	Z	P(Z<z)	P(Z>z)
2	5.27	0.999999932	=1-B2

9. To obtain the remaining cumulative probabilities, copy the formula in cell C2 to cells C3 through C7. A completed table is shown below.

	A	B	C
1	Z	P(Z<z)	P(Z>z)
2	5.27	0.999999932	6.82119E-08
3	2.6725	0.996235581	0.003764419
4	1	0.841344746	0.158655254
5	-1	0.158655254	0.841344746
6	-2.6725	0.003764419	0.996235581
7	-5.27	6.82119E-08	0.999999932

Obtaining z Values for Specified Probabilities Using the NORM.S.INV Function

Researchers sometimes want to know the z value that defines a particular point in the distribution, such as the lower 10% or the upper 5%. To obtain these values, we can use the NORM.S.INV function. The procedure is quite straightforward, so I have selected only four values to illustrate the use of the NORM.S.INV function.

1. Begin by preparing an output table like the one shown at the top of the next page to display the probabilities and associated z values.

	A	B
1	Probability	z
2	Lower 1%	
3	Lower 5%	
4	Upper 5%	
5	Upper 1%	

2. What z value defines the lower 1% of the standard normal distribution? Activate cell **B2** where the z value will be placed.

3. Click the **Formulas** tab near the top of the screen and select **Insert Function**.

4. In the Insert Function dialog box, select the **Statistical** category and the **NORM.S.INV** function. Click **OK**.

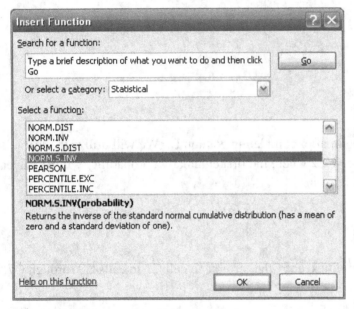

5. Probabilities are represented as proportions. For the lower 1%, enter **.01** in the Probability window as shown below.

6. Click **OK** and a *z* value of -2.3264 is returned.

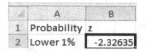

7. What *z* value defines the lower 5% of the standard normal distribution? Activate cell **B3** where the *z* value will be placed.

8. Select **Insert Function** in the Formulas ribbon.

9. In the Insert Function dialog box, select the **Statistical** category and the **NORM.S.INV** function. Click **OK**.

10. For the lower 5%, enter **.05** in the Probability window.

11. Click **OK** and a value of -1.6449 is returned.

12. What *z* value defines the upper 5% of the standard normal distribution? Activate cell **B4** where the *z* value will be placed.

13. Select **Insert Function** in the Formulas ribbon.

14. In the Insert Function dialog box, select the **Statistical** category and the **NORM.S.INV** function. Click **OK**.

15. The point in the normal distribution that defines the upper 5% is the same point that defines the lower 95%. The NORM.S.INV is a cumulative probability function, so we need to enter .95 to obtain the correct *z* value. Click in the Probability window and enter **.95**.

Function Arguments   ? ✕

NORM.S.INV

        Probability   .95   = 0.95

                            = 1.644853627

Returns the inverse of the standard normal cumulative distribution (has a mean of zero and a standard deviation of one).

             Probability   is a probability corresponding to the normal distribution, a number between 0 and 1 inclusive.

Formula result = 1.644853627

Help on this function   [OK]   [Cancel]

16. Click **OK** and a value of 1.6449 is returned.

17. What *z* value defines the upper 1% of the standard normal distribution? Activate cell **B5** where the *z* value will be placed.

18. Select **Insert Function** in the Formulas ribbon.

19. In the Insert Function dialog box, select the **Statistical** category and the **NORM.S.INV** function. Click **OK**.

20. The point in the distribution that defines the upper 1% also defines the lower 99%. Since NORM.S.INV is a cumulative probability function, we need to enter .99 to obtain the correct z value. Click in the **Probability** window and enter **.99**.

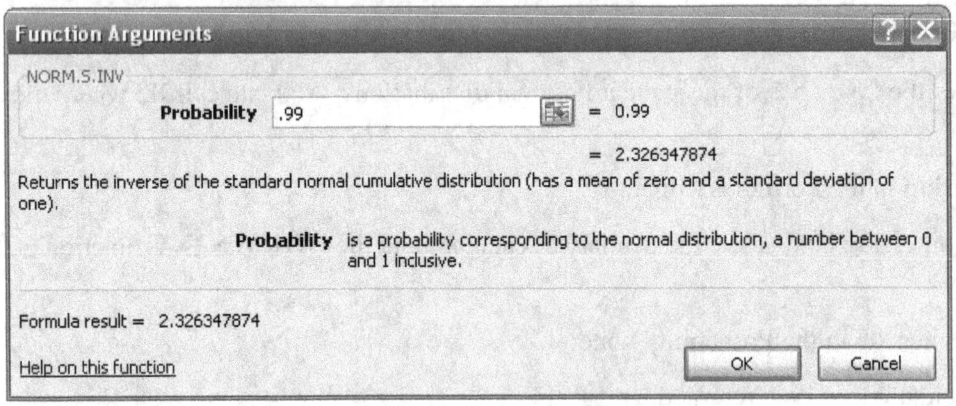

21. Click **OK** and a value of 2.3264 is returned.

A completed output table is shown below.

	A	B
1	Probability	z
2	Lower 1%	-2.32635
3	Lower 5%	-1.64485
4	Upper 5%	1.644854
5	Upper 1%	2.326348

t Distribution

In this chapter, I will limit my instructions regarding Excel's *t* distribution functions to the tasks of obtaining probabilities associated with specified *t* values and obtaining *t* values for specified probabilities. In other words, I will explain how to use Excel's functions to get information that traditionally has been found in tables in statistics textbooks. I think you will agree with me that these functions are much better than the tables. The functions explained in this chapter are listed below.

T.DIST.2T returns the two-tailed Student's t-distribution.

T.DIST.RT returns the right-tailed Student's t-distribution.

T.INV returns the left-tailed inverse of the Student's t-distribution.

T.INV.2T returns the two-tailed inverse of the Student's t distribution.

Obtaining 2-Tailed Probabilities for Specified *t* Values Using the T.DIST.2T Function

1. Let's find the 2-tailed probability of *t* = 1.655 in a distribution with df = 30. Open a new, blank Excel worksheet and click in cell **A1**.

2. Select **Insert Function** in the Formulas ribbon.

3. In the Insert Function dialog box, select the **Statistical** category and the **T.DIST.2T** function as shown at the top of the next page. Click **OK**.

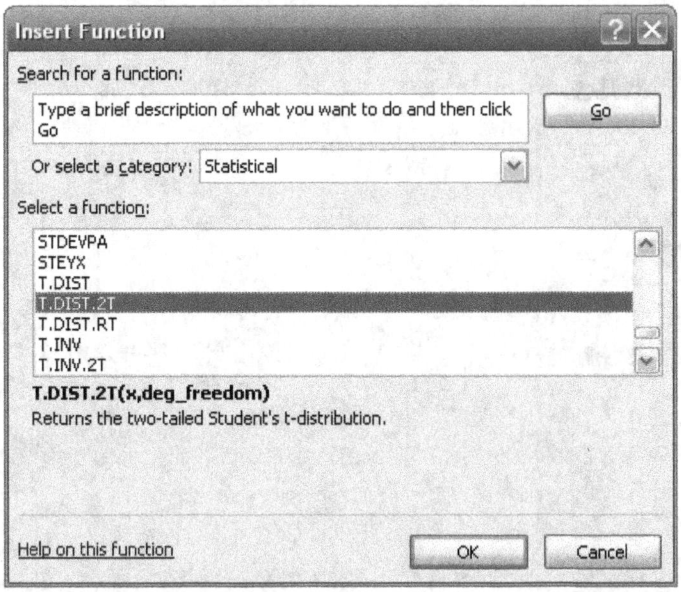

Insert Function

Search for a function:

Type a brief description of what you want to do and then click Go | Go |

Or select a category: | Statistical |

Select a function:

STDEVPA
STEYX
T.DIST
T.DIST.2T
T.DIST.RT
T.INV
T.INV.2T

T.DIST.2T(x,deg_freedom)
Returns the two-tailed Student's t-distribution.

Help on this function | OK | | Cancel |

4. Complete the T.DIST.2T dialog box as shown below.

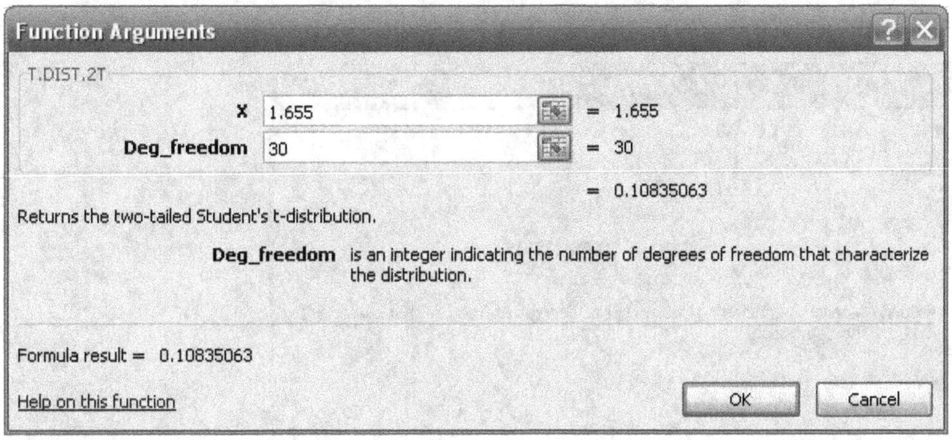

Function Arguments

T.DIST.2T

X | 1.655 | = 1.655
Deg_freedom | 30 | = 30

= 0.10835063

Returns the two-tailed Student's t-distribution.

Deg_freedom is an integer indicating the number of degrees of freedom that characterize the distribution.

Formula result = 0.10835063

Help on this function | OK | | Cancel |

5. Click **OK** and a 2-tailed probability of 0.1084 is returned.

A
1	0.108351

Obtaining 1-Tailed Probabilities for Specified *t* Values Using the T.DIST.RT Function

1. Let's find the 1-tailed probability of $t =1.655$ in a distribution with df $= 30$.

2. The output can be placed in any cell. I selected **A3**.

3. Select **Insert Function** in the Formulas ribbon.

4. In the Insert Function dialog box, select the **Statistical** category and the **T.DIST.RT** function as shown at the top of the next page. Click **OK**.

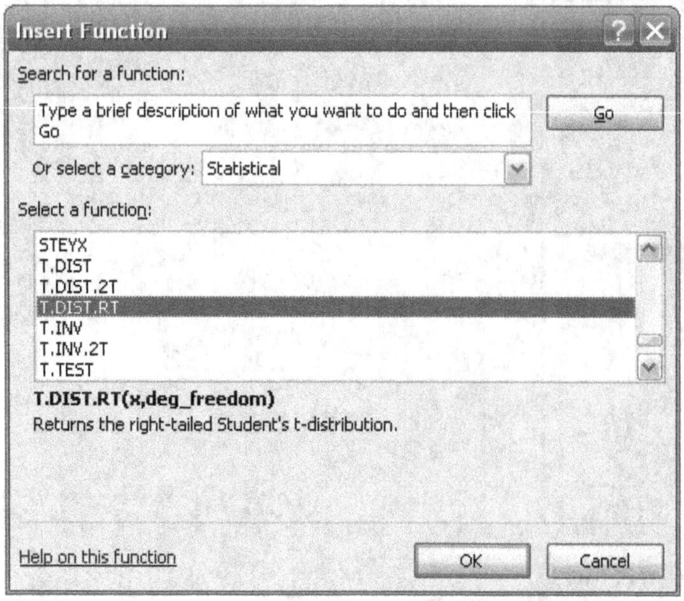

Insert Function   ? X

Search for a function:

| Type a brief description of what you want to do and then click Go | Go |

Or select a category: Statistical

Select a function:

```
STEYX
T.DIST
T.DIST.2T
T.DIST.RT
T.INV
T.INV.2T
T.TEST
```

T.DIST.RT(x,deg_freedom)
Returns the right-tailed Student's t-distribution.

Help on this function   OK   Cancel

5. Complete the T.DIST.RT dialog box as shown below.

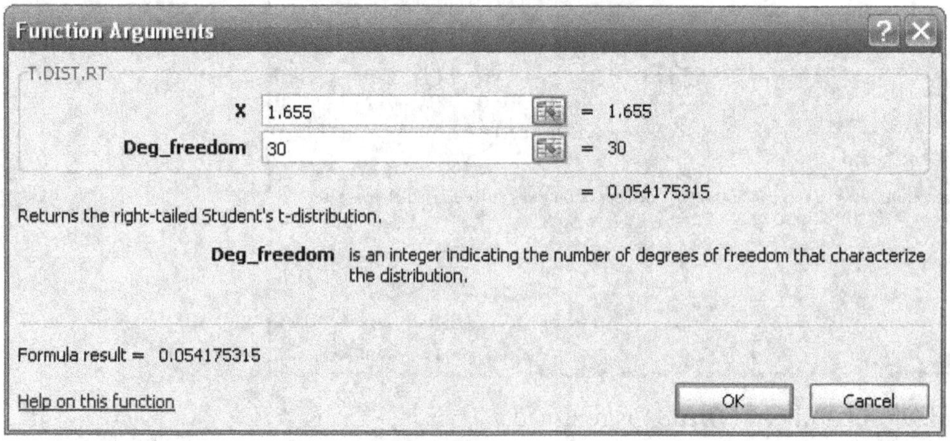

Function Arguments   ? X

T.DIST.RT

X	1.655		= 1.655
Deg_freedom	30		= 30

= 0.054175315

Returns the right-tailed Student's t-distribution.

Deg_freedom   is an integer indicating the number of degrees of freedom that characterize the distribution.

Formula result = 0.054175315

Help on this function   OK   Cancel

6. Click **OK** and 1-tailed probability of 0.0542 is returned. Note that the 1-tailed probability of -1.655 is the same as the 1-tailed probability of 1.655. They are both equal to 0.0542.

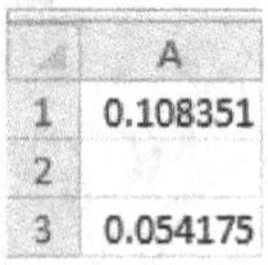

	A
1	0.108351
2	
3	0.054175

Obtaining Critical *t* Values for a Specified 1-Tailed Probabilities Using the T.INV Function

1. Let's find the critical *t* value in a distribution with df = 18 for a 1-tailed probability of 0.05. The output can be placed in any cell. I selected **A1**.

2. Select **Insert Function** in the Formulas ribbon.

3. In the Insert Function dialog box, select the **Statistical** category and the **T.INV** function as shown at the top of the next page. Click **OK**.

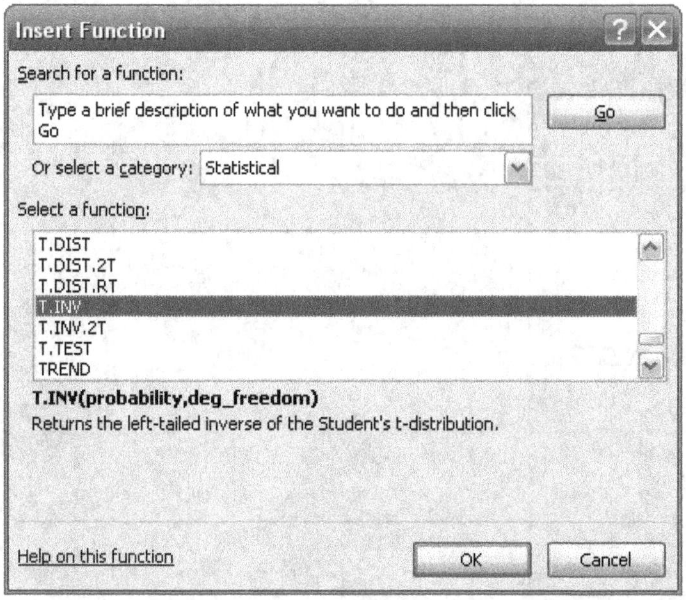

Insert Function   ? ✕

Search for a function:

Type a brief description of what you want to do and then click Go   Go

Or select a category: Statistical ▾

Select a function:

> T.DIST
> T.DIST.2T
> T.DIST.RT
> **T.INV**
> T.INV.2T
> T.TEST
> TREND

T.INV(probability,deg_freedom)
Returns the left-tailed inverse of the Student's t-distribution.

Help on this function   OK   Cancel

4. Complete the T.INV dialog box as shown below.

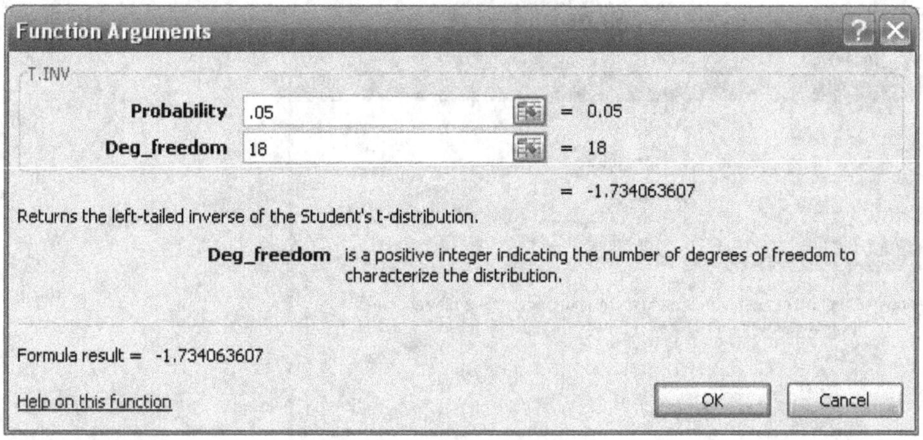

Function Arguments   ? ✕

T.INV

Probability   .05   = 0.05
Deg_freedom   18   = 18

= -1.734063607

Returns the left-tailed inverse of the Student's t-distribution.

Deg_freedom   is a positive integer indicating the number of degrees of freedom to characterize the distribution.

Formula result = -1.734063607

Help on this function   OK   Cancel

5. Click **OK** and a critical t of -1.7341 is returned. Note that the critical t in the right tail of a distribution with df = 18 would be 1.7341.

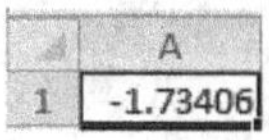

	A
1	-1.73406

Obtaining Critical t Values for Specified 2-Tailed Probabilities Using the T.INV.2T Function

1. Let's find the critical t value in a distribution with df = 18 for a 2-tailed probability of 0.05. I clicked in cell **A3** to place the output there.

2. Select **Insert Function** in the Formulas ribbon.

3. In the Insert Function dialog box, select the **Statistical** category and the **T.INV.2T** function as shown at the top of the next page. Click **OK**.

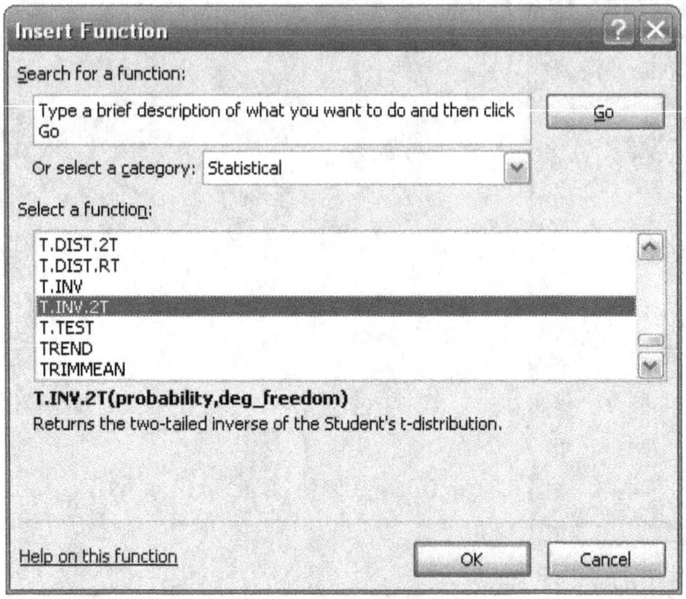

Insert Function

Search for a function:

Type a brief description of what you want to do and then click Go — [Go]

Or select a category: Statistical

Select a function:

```
T.DIST.2T
T.DIST.RT
T.INV
T.INV.2T
T.TEST
TREND
TRIMMEAN
```

T.INV.2T(probability,deg_freedom)
Returns the two-tailed inverse of the Student's t-distribution.

Help on this function   [OK]   [Cancel]

4. Complete the T.INV.2T dialog box as shown below.

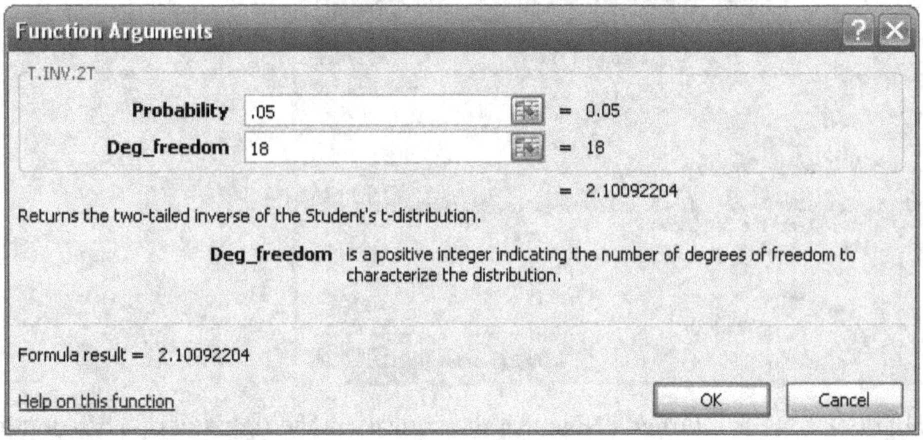

Function Arguments

T.INV.2T

Probability   .05   = 0.05
Deg_freedom   18   = 18

                           = 2.10092204

Returns the two-tailed inverse of the Student's t-distribution.

Deg_freedom is a positive integer indicating the number of degrees of freedom to characterize the distribution.

Formula result = 2.10092204

Help on this function   [OK]   [Cancel]

5. Click **OK** and a critical t value of 2.1009 is returned.

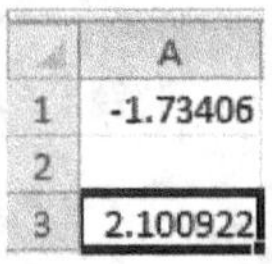

	A
1	-1.73406
2	
3	2.100922

F Distribution

I will first explain how to find probabilities associated with specified *F* values using the F.DIST.RT function. Then I will explain how to obtain critical *F* values for specified probabilities using the F.INV function.

Obtaining Probabilities for Specified *F* Values Using the F.DIST.RT Function

1. Let's find the probability of $F = 3.125$ in a distribution with numerator df = 1 and denominator df = 12. Activate cell A1 where the probability will be placed.

2. Select **Insert Function** in the Formulas ribbon.

3. In the Insert Function dialog box, select the **Statistical** category and the **F.DIST.RT** function. Click
    OK.

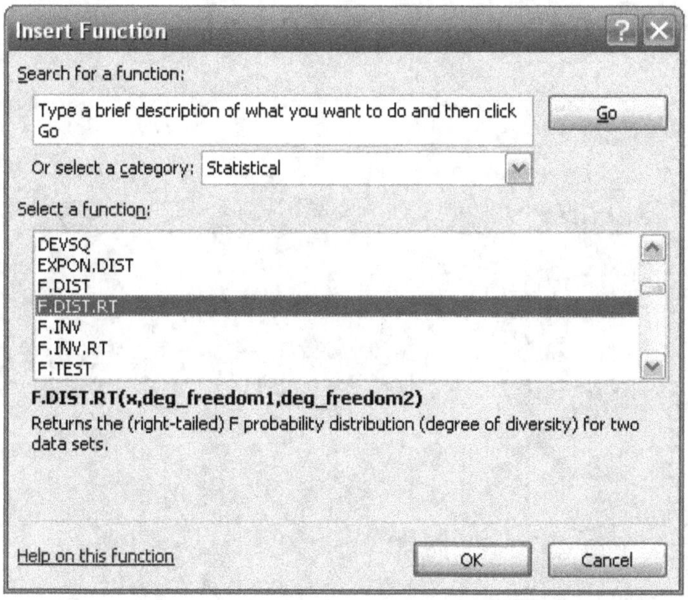

Insert Function

Search for a function:

Type a brief description of what you want to do and then click
Go

[Go]

Or select a category: Statistical

Select a function:

DEVSQ
EXPON.DIST
F.DIST
F.DIST.RT
F.INV
F.INV.RT
F.TEST

F.DIST.RT(x,deg_freedom1,deg_freedom2)
Returns the (right-tailed) F probability distribution (degree of diversity) for two
data sets.

Help on this function   [OK]   [Cancel]

4. Complete the F.DIST.RT dialog box as shown below. Deg_freedom1 refers to numerator df, and
    Deg_freedom1 refers to denominator df.

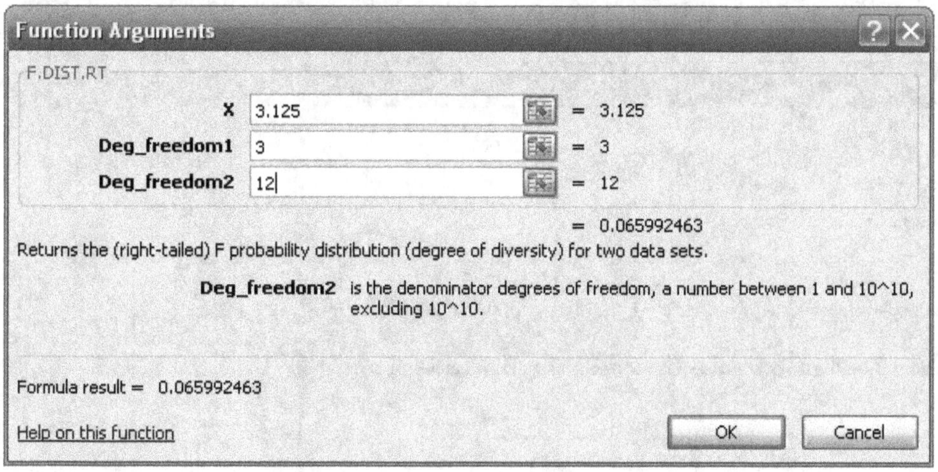

Function Arguments

F.DIST.RT

X	3.125	= 3.125
Deg_freedom1	3	= 3
Deg_freedom2	12	= 12

= 0.065992463

Returns the (right-tailed) F probability distribution (degree of diversity) for two data sets.

  Deg_freedom2 is the denominator degrees of freedom, a number between 1 and 10^10,
            excluding 10^10.

Formula result = 0.065992463

Help on this function   [OK]   [Cancel]

5. Click **OK** and a probability of 0.0660 is returned.

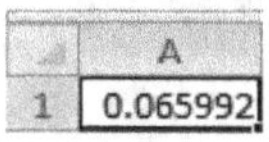

	A
1	0.065992

Obtaining *F* Values for Specified Probabilities Using the F.INV.RT Function

To obtain a critical *F* value, we will use the F.INV.RT function. We need to specify probability (alpha),
numerator df (df 1), and denominator df (df 2).

1. Let's find the critical F for a probability equal to .025 in a distribution with numerator df $= 4$ and denominator df $= 163$. Activate cell **A1** in a blank Excel worksheet.

2. Select **Insert Function** in the Formulas ribbon.

3. In the Insert Function dialog box, select the **Statistical** category and the **F.INV.RT** function. Click **OK**.

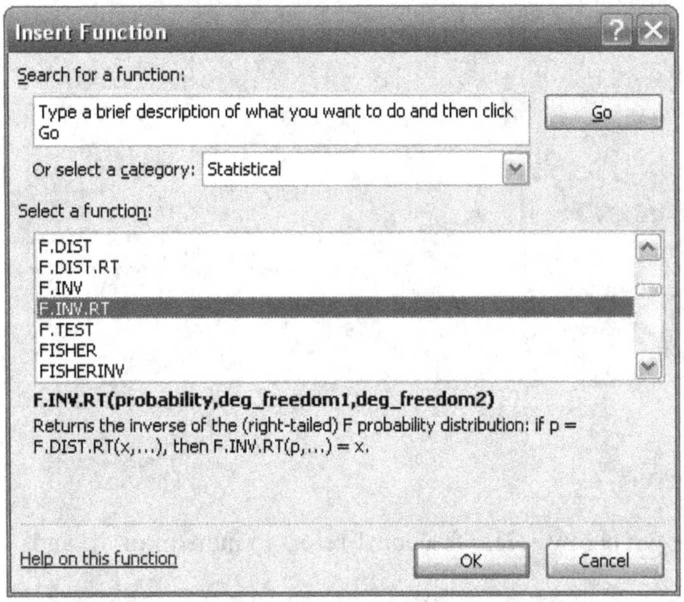

4. Complete the F.INV.RT dialog box as shown below.

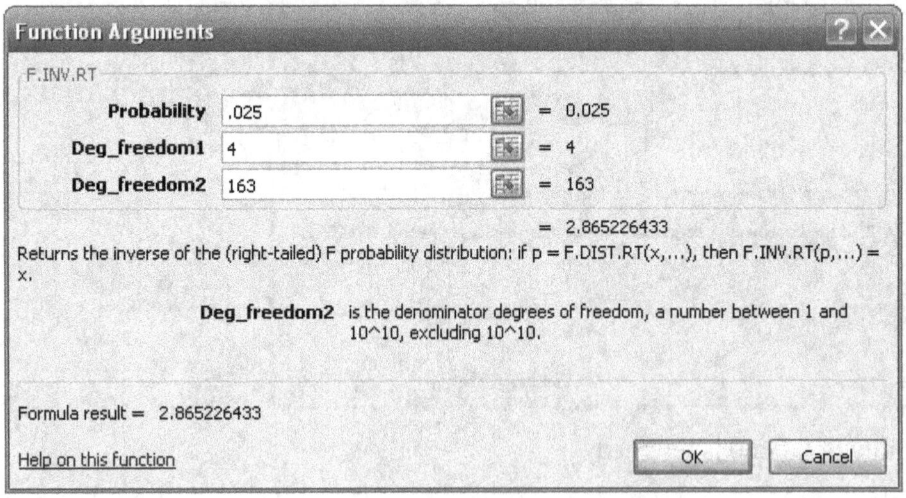

5. Click **OK** and a value of 2.8652 is returned.

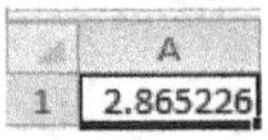

Chi-Square Distribution

I will first explain how to find probabilities associated with specified chi-square values using the CHI.DIST.RT function. Then I will explain how to obtain critical chi-square values for specified probabilities using the CHI.INV.RT function.

Obtaining Probabilities for Specified Chi-Square Values Using the CHISQ.DIST.RT Function

1. Let's find the probability of chi-square equal to 30.5682 in a distribution with 3 degrees of freedom. Click in cell **A1** where the output will be placed.

2. Select **Insert Function** in the Formulas ribbon.

3. In the Insert Function dialog box, select the **Statistical** category and the **CHISQ.DIST.RT** function. Click **OK**.

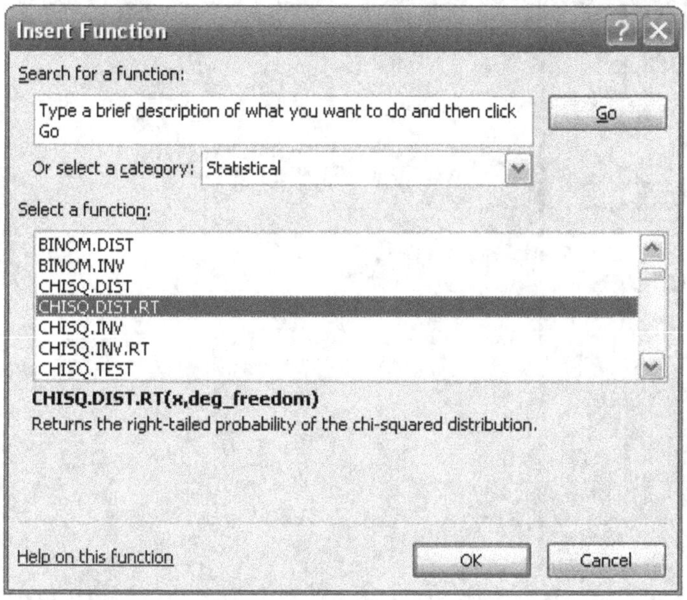

4. Complete the CHISQ.DIST.RT dialog box as shown below.

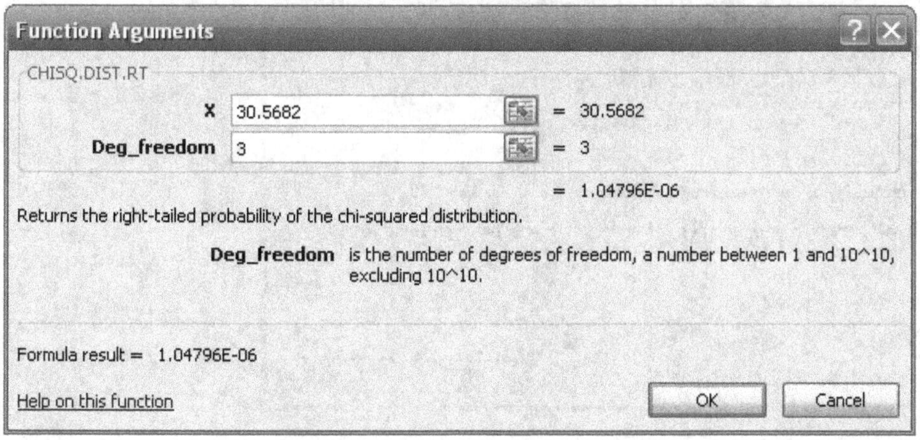

5. Click **OK** and a probability of 1.04796E-06 is returned. Scientific notation is used in the output whenever the probability is very small.

	A
1	1.04796E-06

Obtaining Chi-Square Values for Specified Probabilities Using the CHISQ.INV.RT Function

To obtain a critical chi-square value, we will use the CHISQ.INV.RT function. We need to specify probability (alpha) and degrees of freedom.

1. Let's find the critical chi-square for a probability of 0.01 in a distribution with df = 2. I clicked in cell **A3** to place the output there.

2. Select **Insert Function** in the Formulas ribbon.

3. In the Insert Function dialog box, select the **Statistical** category and the **CHISQ.INV.RT** function. Click **OK**.

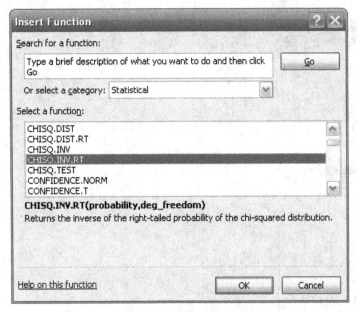

4. Complete the CHISQ.INV.RT dialog box as shown below.

5. Click **OK** and a critical value of 9.2103 is returned.

3	9.210340372

Chapter 7

Testing Hypotheses
About One Sample Means

If a researcher wants to compare the mean of one sample with a hypothesized population value, he or she should use the one-sample z-test or the one-sample t-test. The selection will be based on the assumptions underlying the tests. For either test, the researcher will have one of three objectives in mind:

1. To test whether the sample mean is greater than the hypothesized population value,

2. To test whether the sample mean is less than the hypothesized population value, or

3. To test whether the sample mean is not equal to the hypothesized population value. That is to say, the sample mean could be either greater than or less than the hypothesized value.

Objectives 1 and 2 are associated with directional or one-tailed tests; objective 3 is associated with a nondirectional or two-tailed test. In this chapter, I first explain how to utilize Excel to carry out the one-sample z-test, and then I explain the one-sample t-test. For both the z-test and the t-test, I have included the one-tailed and two-tailed versions. I also describe how to set up confidence intervals for the nondirectional tests.

► Section 7.1 | One-Sample z-Test

The formula for the one-sample z-test is

$$z = \frac{\bar{X} - \mu}{\sigma_{\bar{x}}}$$

where $\bar{X}$ is the sample mean, μ is the hypothesized value of the population mean, and $\sigma_{\bar{x}}$ is the standard error of the mean.

The formula for the standard error of the mean is given by

$$\sigma_{\bar{x}} = \frac{\sigma}{\sqrt{n}}$$

where σ is the standard deviation of the population, and n is the number of subjects in the sample.

Assumptions Underlying the z-Test

The statistical assumptions underlying the one-sample z-test include:

1. Observations are independent of one another.

2. The observations are randomly sampled from the population.

3. Observations are normally distributed in the population.

4. The population variance, σ^2, is known.

Sample Research Problem

The increase in the number of children who are schooled at home has stimulated numerous research questions regarding the characteristics of these children. One such question is related to their level of intelligence. Do home-schooled children possess the same average level of intelligence as other children?   It might be assumed that parents who decide to educate their children at home are fairly bright, and that their children, in turn, are also fairly bright. A researcher who was interested in this question randomly selected 12 seven-year-olds who were being home-schooled and administered to each child the Wechsler Intelligence Scale for Children (WISC). The published norms for the WISC indicate that the scores for the population are normally distributed with a mean equal to 100 and a standard deviation equal to 15. Is the mean score on the WISC for seven-year-old home-schooled children significantly different from 100?

Steps to Follow to Analyze the Sample Research Problem

1. Open worksheet "Ch7_WISC" on the Web site, or enter the data and output table in an Excel worksheet as shown below. Be sure to use exactly the same cell locations as I have, because I frequently refer to cell locations in the instructions. You may need to make column C wider to accommodate the long labels.

	A	B	C	D
1	WISC		ONE-SAMPLE Z-TEST	
2	97		Sample mean	
3	112		Hypothesized population mean	100
4	119		Population standard deviation	15
5	84		Count	
6	135		Standard error of the mean	
7	95		Z	
8	127		Alpha	0.05
9	103		Probability one-tailed	
10	95		Z critical one-tailed	
11	107		Probability two-tailed	
12	101		Z critical two-tailed	
13	98			
14			CONFIDENCE INTERVAL	
15			Lower Limit	
16			Upper Limit	

2. Use the AVERAGE function to obtain the sample mean. Activate cell **D2**. Click the **Formulas** tab near the top of the screen and select **Insert Function**.

3. In the Insert Function dialog box, select the **Statistical** category. Select **AVERAGE** in the function list. Click **OK**.

```
Insert Function                                            ? X

Search for a function:

Type a brief description of what you want to do and then click    [  Go  ]
Go

Or select a category:  Statistical                    v

Select a function:

AVEDEV                                                  ^
AVERAGE
AVERAGEA
AVERAGEIF
AVERAGEIFS
BETA.DIST
BETA.INV                                                v

AVERAGE(number1,number2,...)
Returns the average (arithmetic mean) of its arguments, which can be numbers or
names, arrays, or references that contain numbers.


Help on this function               [   OK   ]    [ Cancel ]
```

4. You now need to tell Excel where the data are located. Click in the **Number 1** window and remove the information, if any, that appears there. Then click in the top cell of the WISC data (cell **A2**), and drag to the end of the data (cell **A13**). If you prefer, you can manually enter **A2:A13** in the Number 1 window.

The AVERAGE function will also give you an accurate result if you include cell A1 (WISC) in the Number 1 range.

```
Function Arguments                                          ? X

AVERAGE

     Number1  [ A2:A13                ]  [ ]  =  {97;112;119;84;135;95;127;103;95;...
     Number2  [                       ]  [ ]  =  number



                                          =  106.0833333
Returns the average (arithmetic mean) of its arguments, which can be numbers or names, arrays, or references that
contain numbers.

              Number1:  number1,number2,... are 1 to 255 numeric arguments for which you want the
                        average.


Formula result =  106.0833333

Help on this function                    [   OK   ]    [ Cancel ]
```

5. Click **OK** and 106.0833 will appear in cell D2 of your worksheet.

6. Next you will use the COUNT function to count the number of observations in the sample. Activate cell **D5**. Select **Insert Function** in the Formulas ribbon.

7. In the Insert Function dialog box, select the **Statistical** category. Select **COUNT** in the function list. Click **OK**.

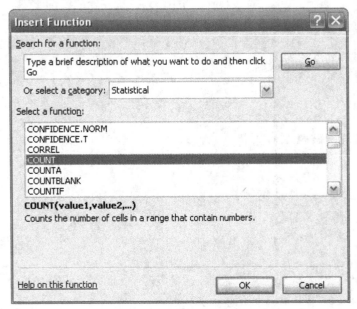

8. Delete information, if any, that appears in the Value 1 window. Then click in the **Value 1** window, click in the top cell of WISC data (**A2**), and drag to the end of the data (**A13**). If you prefer, you can manually enter **A2:A13** in the Value 1 window.

9. Click **OK** and 12 will appear in cell D5 of the worksheet.

10. **Standard error of the mean**. The formula for the standard error of the mean is

$$\sigma_{\bar{x}} = \frac{\sigma}{\sqrt{n}}$$

You will use this formula to calculate the standard error. Activate cell **D6** where you will place the value of the standard error.

The formula you will be entering in cell D6 is =D4/SQRT(D5). The equal sign tells Excel that the information that follows will be a formula. D4 is the cell location of σ, and D5 is the cell location of n (Count). The diagonal indicates division. SQRT is the abbreviation for the square root function. Excel will compute the square root of the number in parentheses immediately following SQRT.

11. Enter **=D4/SQRT(D5)** in cell D6. Press **[Enter]** and a value of 4.3301 will be returned.

	A	B	C	D	E
1	WISC		ONE-SAMPLE Z-TEST		
2	97		Sample mean	106.0833	
3	112		Hypothesized population mean	100	
4	119		Population standard deviation	15	
5	84		Count	12	
6	135		Standard error of the mean	=D4/SQRT(D5)	

12. You will now key in another formula to calculate *z*. Recall that the formula for *z* is

$$z = \frac{\overline{X} - \mu}{\sigma_{\overline{x}}}$$

Activate cell **D7** and enter **=(D2-D3)/D6**.

The period after D6 is not part of the formula.

	A	B	C	D	E
1	WISC		ONE-SAMPLE Z-TEST		
2	97		Sample mean	106.0833	
3	112		Hypothesized population mean	100	
4	119		Population standard deviation	15	
5	84		Count	12	
6	135		Standard error of the mean	4.330127	
7	95		Z	=(D2-D3)/D6	

13. Press **[Enter]** and 1.4049 will appear in cell D7 of the worksheet. This is the obtained value of *z*.

14. Next, you will find the one-tailed probability associated with the obtained value of *z*. Activate cell **D9** where the one-tailed probability will be displayed. You will be using a formula to obtain the one-tailed probability. When completed, it will look like:

=1-NORM.S.DIST(ABS(D7))

NORM.S.DIST is the standard normal cumulative distribution function. It returns the cumulative probability associated with a given z value. ABS is the absolute value function. Because it is possible for z to take on negative as well as positive values, the task of evaluating the statistical significance of z will be simplified if you are working with an absolute value. Finally, D7 is the cell location of z.

15.   Enter **=1-** and then select **Insert Function** in the Formulas ribbon.

7	95	z	1.404886
8	127	Alpha	0.05
9	103	Probability one-tailed	=1-

16.   In the Insert Function dialog box, select the **Statistical** category. Select **NORM.S.DIST** in the function list. Click **OK**.

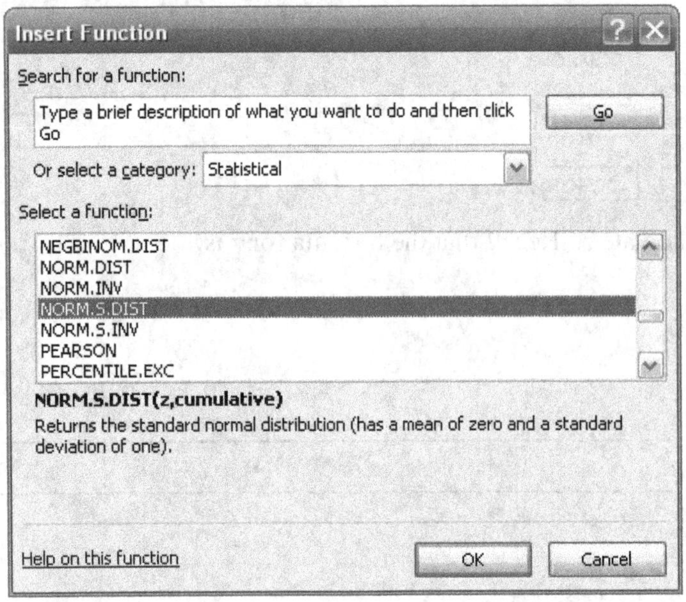

17.   Enter **ABS(D7)** in the *z* window and click **OK**. A value of .0800, the one-tailed probability of obtained *z*, will appear in cell D9 of the worksheet.

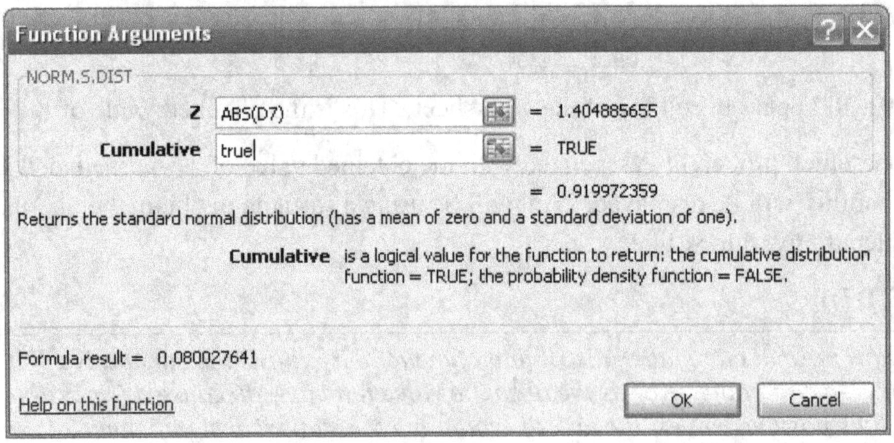

18. Next, you will obtain **z critical one-tailed**. Activate cell **D10**. The formula that you will be entering is:

=ABS(NORM.S.INV(.05))

This formula contains two functions. ABS is the absolute value function. NORM.S.INV is the inverse of the standard normal cumulative distribution function. NORM.S.INV provides the z value associated with a specified probability. You want the z value associated with a probability of .05, the selected value of alpha.

---

19. Key in the formula **=ABS(NORM.S.INV(.05))** and press [**Enter**]. A critical *z* of 1.6449 will be displayed in the output table.

| 9 | 103 | Probability one-tailed | 0.080028 |
| 10 | 95 | Z critical one-tailed | =ABS(NORM.S.INV(.05)) |

20. Activate cell **D11** where you will place the two-tailed probability. Because the two-tailed probability can be obtained by doubling the number shown for the one-tailed probability, we will enter a simple formula to compute the two-tailed probability using the number displayed in cell D9. Key in **=2*D9**. Press [**Enter**] and .1601 should appear in your worksheet.

9	103	Probability one-tailed	0.080028
10	95	Z critical one-tailed	1.644854
11	107	Probability two-tailed	=2*D9

21. The last entry in the table is the two-tailed critical value of *z*. To obtain this value, you will enter a formula like the one you used for the one-tailed critical value, with one exception. Alpha is equally divided between the upper and lower tails for a two-tailed test, so you will use .025 in the formula instead of .05. Activate **D12**. Key in the formula shown below.

=ABS(NORM.S.INV(.025))

| 11 | 107 | Probability two-tailed | 0.160055 |
| 12 | 101 | Z critical two-tailed | =ABS(NORM.S.INV(.025)) |

22. Press [**Enter**] and 1.959964 will be displayed in the table.

Interpreting the Output

	A	B	C	D
1	WISC		ONE-SAMPLE Z-TEST	
2	97		Sample mean	106.0833
3	112		Hypothesized population mean	100
4	119		Population standard deviation	15
5	84		Count	12
6	135		Standard error of the mean	4.330127
7	95		Z	1.404886
8	127		Alpha	0.05
9	103		Probability one-tailed	0.080028
10	95		Z critical one-tailed	1.644854
11	107		Probability two-tailed	0.160055
12	101		Z critical two-tailed	1.959964
13	98			

- **Sample mean.** The mean WISC score for the 12 children in the sample, $\bar{X} = 106.0833$.

- **Hypothesized population mean.** The value 100 indicates that the hypothesis being tested in this analysis is that the population mean of children who are home-schooled is equal to 100.

- **Population standard deviation.** The population standard deviation (or the population variance) must be known in order for the z-test to be applied appropriately to a set of data. For this analysis, the population standard deviation is equal to 15.

- **Count.** The number of observations in the sample *(n)*. For this problem, n is equal to 12.

- **Standard error of the mean.** The standard error is the standard deviation of the sampling distribution of the mean, and, as such, it provides an approximation of the average amount by which the sample means deviate from the population mean. For this example, you can say that the distribution of sample means for samples of size 12 drawn from the population of WISC scores for home-schooled children has a standard deviation of 4.3301 points.

- **z.** Value of z obtained by applying the formula presented at the beginning of this section. For this research problem, obtained z equals 1.4049.

- **Alpha.** The significance level for the statistical test. This value is selected by the researcher(s), and for this analysis, we selected .05.

- **Probability one-tailed.** One-tailed probability of the obtained z, also referred to as the one-tailed P-value. For this problem, if the null hypothesis is true that the population mean WISC score for home-schooled children is 100, then the chance probability of obtaining a $z \leq -1.4049$ is .0800. Or, in the upper tail of the distribution, the chance probability of obtaining a $z \geq 1.4049$ is .0800. Because the probability of obtained z is greater than alpha, you would conclude that the result is not statistically significant with a one-tailed test.

- **z critical one-tailed.** Absolute value of the one-tailed critical value of z for the selected value of alpha. With alpha equal to .05, and an alternative hypothesis that says that the population mean WISC score for home-schooled children is less than 100, the one-tailed critical value of z is -1.6445. Similarly, for an alternative hypothesis that says that the population mean WISC score for home-schooled children is greater than 100, the one-tailed critical value of z is 1.6445. Assuming that your alternative hypothesis stated that the population mean was greater than 100, you would have to conclude that the one-tailed test is not statistically significant because the obtained z of 1.4049 is not greater than the critical z.

- **Probability two-tailed.** Two-tailed chance probability (or two-tailed P-value) associated with the obtained z. More specifically, if the null hypothesis is true that the population mean WISC score for home-schooled children is 100, then the chance probability of obtaining a $z \leq -1.4049$ plus the chance probability of obtaining a $z \geq 1.4049$ is .1601. Note that this value is two times the one-tailed probability of .0800. Because .1601 is greater than alpha (.05), you would conclude that the two-tailed test is not statistically significant.

- **z critical two-tailed.** Absolute value of the two-tailed critical value of z for the selected value of alpha. For alpha equal to .05, the two-tailed critical value of z is 1.9600. This result is not statistically significant, because obtained z (1.4049) is less than critical z.

Confidence Interval for the One-Sample z-Test

To calculate the lower and upper limits of the confidence interval, we use the following general expression:

$$\bar{X} \pm (z_{Crit})(\sigma_{\bar{x}})$$

where $\bar{X}$ represents the sample mean, z_{Crit} represents the two-tailed z critical value, and $\sigma_{\bar{x}}$ represents the standard error of the mean. We will be using information that is provided in the ONE-SAMPLE Z-TEST output table. For easy reference, let's place the confidence interval output just below that table.

1. Key in the following labels: **CONFIDENCE INTERVAL**, **Lower limit**, and **Upper limit**. Be sure to use the same cell placement as shown below, because I refer to these cell locations in the instructions.

11	107	Probability two-tailed	0.160055
12	101	Z critical two-tailed	1.959964
13	98		
14		CONFIDENCE INTERVAL	
15		Lower Limit	
16		Upper Limit	

2. We will now compute the lower limit. Activate cell **D15**. Enter **=D2-(D12*D6)**.

In this formula, D2 is the cell location of the sample mean, D12 is the cell location of the two-tailed Z critical value, * indicates multiplication, and D6 is the cell location of the standard error of the mean.

| 14 | CONFIDENCE INTERVAL | |
| 15 | Lower Limit | =D2-(D12*D6) |

3. Press [**Enter**]. You should now see 97.5964 in cell D15 of the output table.

4. The formula for the upper limit is nearly identical to the formula for the lower limit, except that you will be adding instead of subtracting. Activate cell **D16** and enter **=D2+(D12*D6)**.

14	CONFIDENCE INTERVAL	
15	Lower Limit	97.59644
16	Upper Limit	=D2+(D12*D6)

5. Press [**Enter**] and 114.5702 should appear in cell D16 of the output table.

14	CONFIDENCE INTERVAL	
15	Lower Limit	97.59644
16	Upper Limit	114.5702

Interpreting the Confidence Interval

The confidence interval for a population mean specifies the range of values that, with a known degree of confidence, includes the unknown population mean. The computing formula for the confidence interval contains two elements: an estimate of the unknown population mean, and estimation error. If the test result is statistically significant, then the sample mean provides a good estimate of the unknown population mean. Estimation error is represented by the product, $(z_{Crit})(\sigma_{\bar{x}})$. When alpha is equal to .05, a 95% confidence interval will be constructed. The 95% confidence interval for our research problem would take the form: $97.5964 \le \mu \le 114.5702$. Based on this interval, we might claim that we are 95% confident that the true population mean lies between 97.5964 and 114.5702. For a confidence level interpretation, we would say that, if we repeated the data analysis many times, sampling all possible samples of n equals 12 children from the population of home-schooled children, 95% of the confidence intervals would include the true population mean. The remaining 5% would not. Note that the confidence interval is consistent with an obtained z-test value that was not statistically significant, because the hypothesized value of the population mean (100) is included in the interval.

▶ Section 7.2 One-Sample *t*-Test

The formula for the one-sample *t*-test is given by

$$t = \frac{\overline{X} - \mu}{S_{\overline{X}}}$$

where $\overline{X}$ is the sample mean, μ is the hypothesized value of the population mean, and $S_{\overline{X}}$ is the estimated standard error of the mean.

The formula for the estimated standard error of the mean is given by

$$S_{\overline{X}} = \frac{S}{\sqrt{n}}$$

where S is the standard deviation of the sample and n is the number of observations included in the sample.

Assumptions Underlying the *t*-Test

The statistical assumptions underlying the one-sample *t*-test are as follows:

1. Observations are independent of one another.
2. The observations are randomly sampled from the population.
3. Observations are normally distributed in the population.
4. The population variance, σ^2, is not known.

Sample Research Problem

It was recently reported that adults in the United States spend an average of 28 hours each week watching TV. A media researcher wanted to find out if adults in other countries also watched TV an average of 28 hours each week. The researcher asked 15 randomly selected Canadian adults to maintain careful records of their TV viewing time for one week.

Steps to Follow to Analyze the Sample Research Problem

1. Open worksheet "Ch7_Hours" on the Web site, or enter the data and output table in an Excel worksheet as shown at the top of the next page. Enter the labels in column C. Enter the numerical information in column D. Be sure to use exactly the same cell locations, because cell addresses appear frequently in

the instructions. You may want to make column C wider so that the longest label will be displayed in its entirety.

	A	B	C	D
1	HOURS		ONE-SAMPLE T-TEST	
2	15		Sample mean	
3	7		Hypothesized population mean	28
4	32		Sample standard deviation	
5	26		Count	
6	31		Standard error of the mean	
7	5		t	
8	0		Alpha	0.01
9	14		df	
10	11		Probability one-tailed	
11	3		t critical one-tailed	
12	38		Probability two-tailed	
13	10		t critical two-tailed	
14	9			
15	6		CONFIDENCE INTERVAL	
16	11		Lower Limit	
17			Upper Limit	

2.   You will obtain the value of the sample mean by using the AVERAGE function. Activate cell **D2**. Click the **Formulas** tab at the top of the screen and select **Insert Function**.

3.   In the Insert Function dialog box, select the **Statistical** category. Select **AVERAGE** in the function list. Click **OK**.

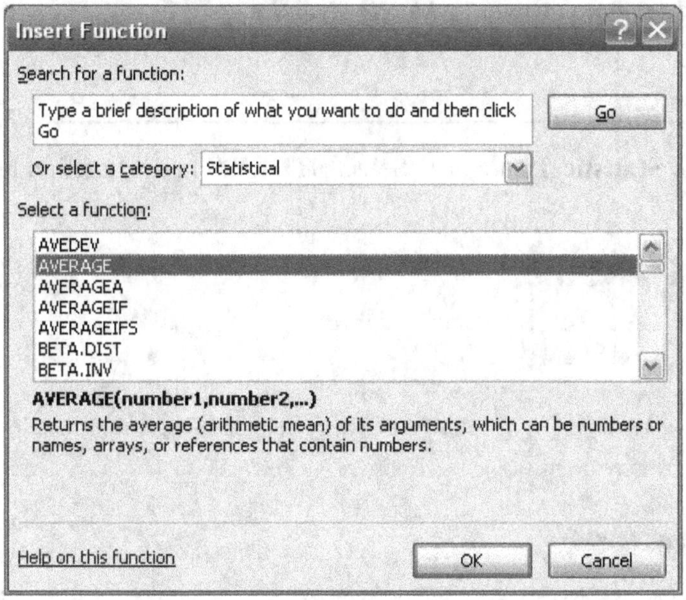

Insert Function

Search for a function:

Type a brief description of what you want to do and then click Go | Go |

Or select a category: Statistical

Select a function:

AVEDEV
AVERAGE
AVERAGEA
AVERAGEIF
AVERAGEIFS
BETA.DIST
BETA.INV

AVERAGE(number1,number2,...)
Returns the average (arithmetic mean) of its arguments, which can be numbers or names, arrays, or references that contain numbers.

Help on this function | OK | Cancel |

4. Delete the information, if any, that appears in the **Number 1** window. Then click in the top cell of the HOURS data (**A2**) and drag to the end of the data (**A16**). If you prefer, you can manually enter **A2:A16** in the Number 1 window.

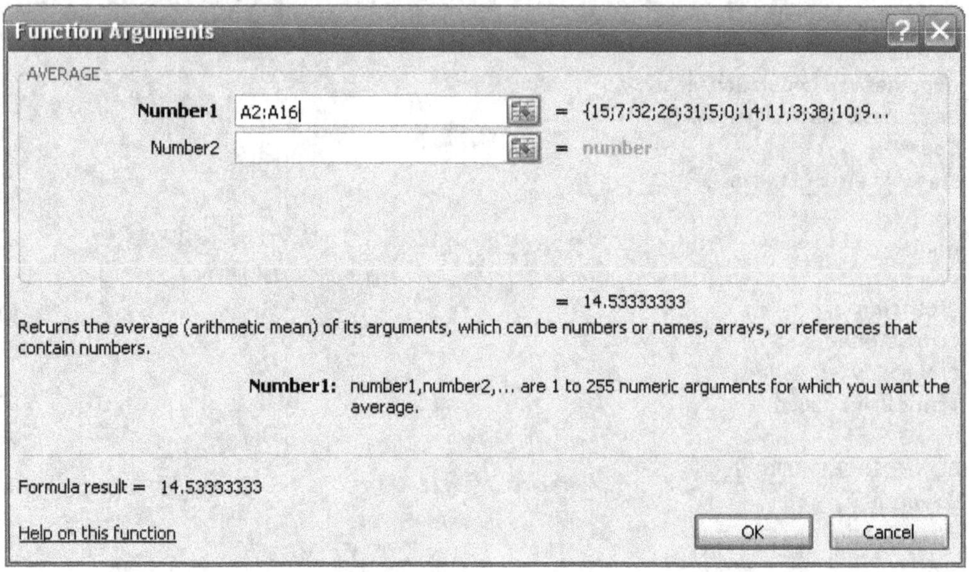

5. Click **OK** and 14.5333 should appear in cell D2 of the worksheet.

6. Use the STDEV.S function to obtain the standard deviation of the sample data. Activate cell **D4**. Click **Insert Function** in the Formulas ribbon.

The STDEV.S function uses the unbiased estimate formula, which is given by

$$S = \sqrt{\frac{\sum (X - \bar{X})^2}{n-1}}$$

---

7. In the Insert Function dialog box, select the **Statistical** category. Select **STDEV.S** in the function list. Click **OK**.

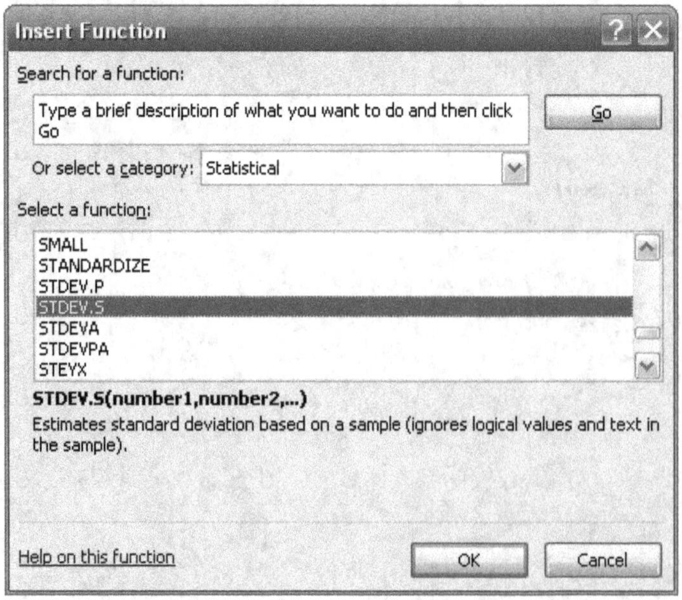

8. Delete information, if any, that appears in the **Number 1** window. Then click in the top cell of the HOURS data (**A2**) and drag to the end of the data (**A16**). If you prefer, you can manually enter **A2:A16** in the Number 1 window.

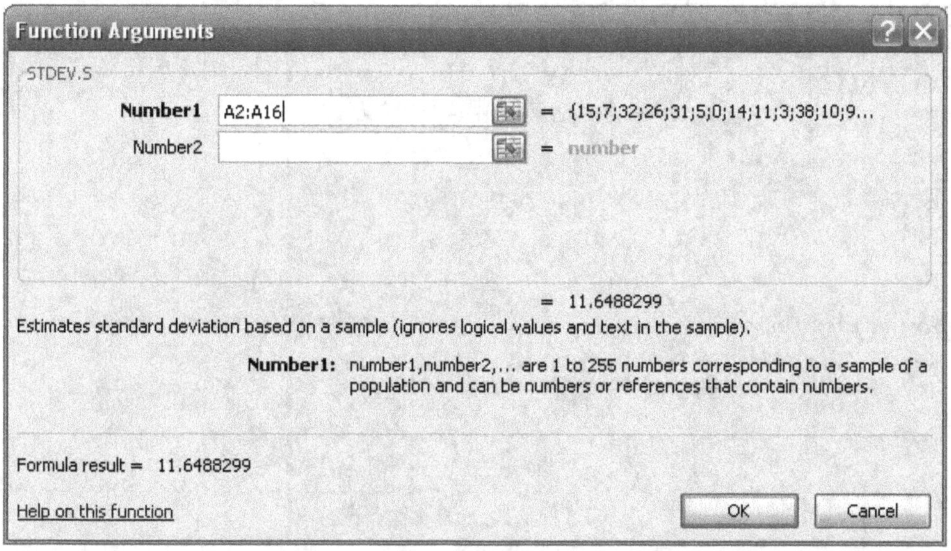

9. Click **OK** and 11.6488 will appear in cell D4 of the output table.

	A	B	C	D
1	HOURS		ONE-SAMPLE T-TEST	
2	15		Sample mean	14.53333
3	7		Hypothesized population mean	28
4	32		Sample standard deviation	11.64883

10. Use the COUNT function to obtain a count of the observations in the sample. Activate cell **D5**. Click **Insert Function** in the Formulas ribbon.

11. In the Insert Function dialog box, select the **Statistical** category. Select **COUNT** in the function list. Click **OK**.

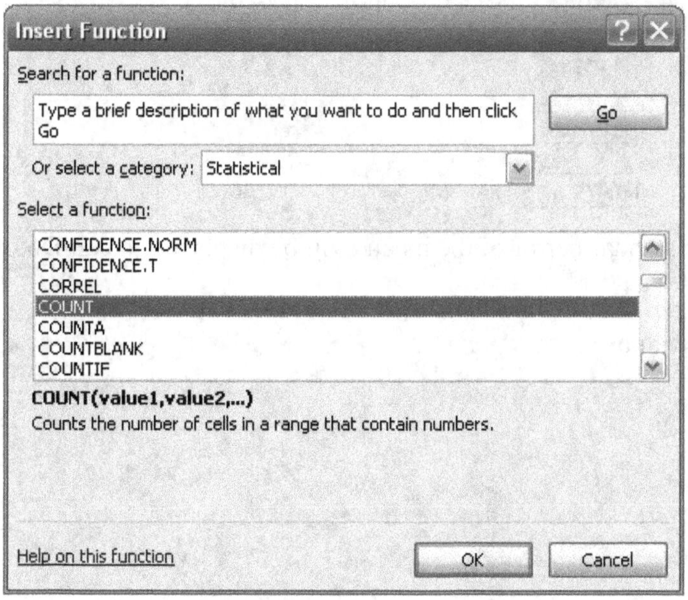

12. Delete information, if any, that appears in the **Value 1** window. Then click in the top cell of the HOURS data (**A2**) and drag to the end of the data (**A16**). If you prefer, you can manually enter **A2:A16** in the Value 1 window.

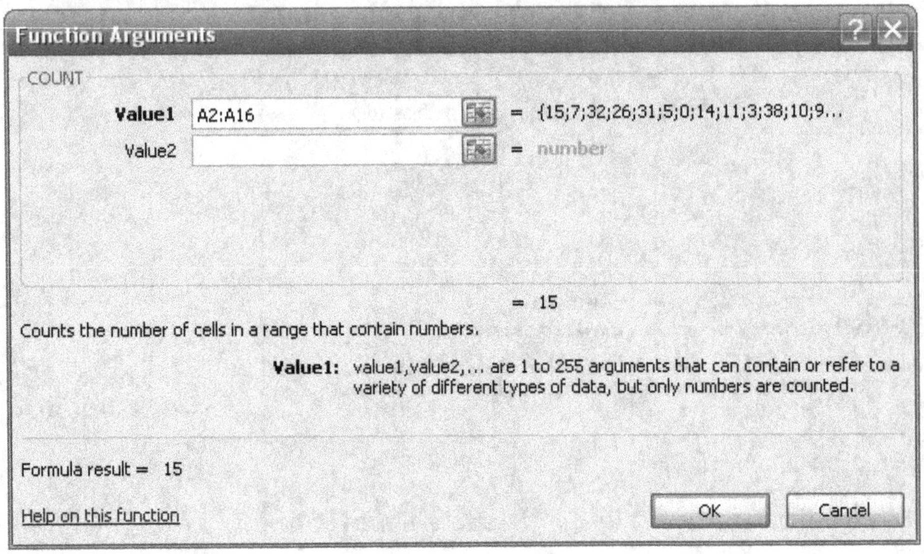

13. Click **OK** and 15 will appear in cell D5 of the worksheet.

14. The formula for the standard error of the mean for the *t*-test is given by

$$S_{\bar{x}} = \frac{S}{\sqrt{n}}$$

Activate cell **D6** in the output table. Key in the formula **=D4/SQRT(D5)**.

The equal sign tells Excel that the information that follows will be a formula. D4 is the cell address of S, the diagonal indicates division, SQRT is the abbreviation for the square root function, and D5 is the cell address of n.

⊿	A	B	C	D	E
1	HOURS		ONE-SAMPLE T-TEST		
2	15		Sample mean	14.53333	
3	7		Hypothesized population mean	28	
4	32		Sample standard deviation	11.64883	
5	26		Count	15	
6	31		Standard error of the mean	=D4/SQRT(D5)	

15. Press [**Enter**] and 3.0077, the value of the standard error of the mean, will be displayed in cell D6 of the output table.

16. The formula for the one-sample *t*-test is given by

$$t = \frac{\bar{X} - \mu}{S_{\bar{x}}}$$

To apply this formula to the data, activate cell **D7** and enter **=(D2-D3)/D6**.

	A	B	C	D	E
1	HOURS		ONE-SAMPLE T-TEST		
2	15		Sample mean	14.53333	
3	7		Hypothesized population mean	28	
4	32		Sample standard deviation	11.64883	
5	26		Count	15	
6	31		Standard error of the mean	3.007715	
7	5		t	=(D2-D3)/D6	

17. Press **[Enter]** and the obtained value of *t*, −4.4774, will appear in cell D7 of the output table.

18. The formula for degrees of freedom (df) for the one-sample *t*-test is $n - 1$.   With *n* equal to 15, df = 14.   Enter **14** in cell **D9** of the output table.

	A	B	C	D
1	HOURS		ONE-SAMPLE T-TEST	
2	15		Sample mean	14.53333
3	7		Hypothesized population mean	28
4	32		Sample standard deviation	11.64883
5	26		Count	15
6	31		Standard error of the mean	3.007715
7	5		t	-4.47737
8	0		Alpha	0.01
9	14		df	14

19. You will use the **T.DIST** function to find the one-tailed probability of obtained t (−4.4774).   Activate cell **D10** in the output table.   Click **Insert Function** in the Formulas ribbon.

20. In the Insert Function dialog box, select the **Statistical** category and the **T.DIST** function.   Click **OK**.

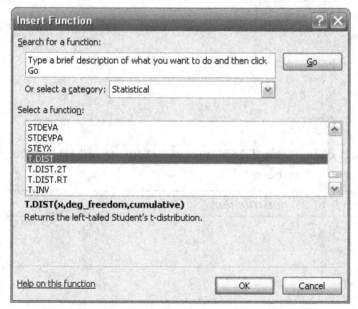

21. Complete the T.DIST dialog box as shown below. An explanation of the entries is given immediately after the dialog box.

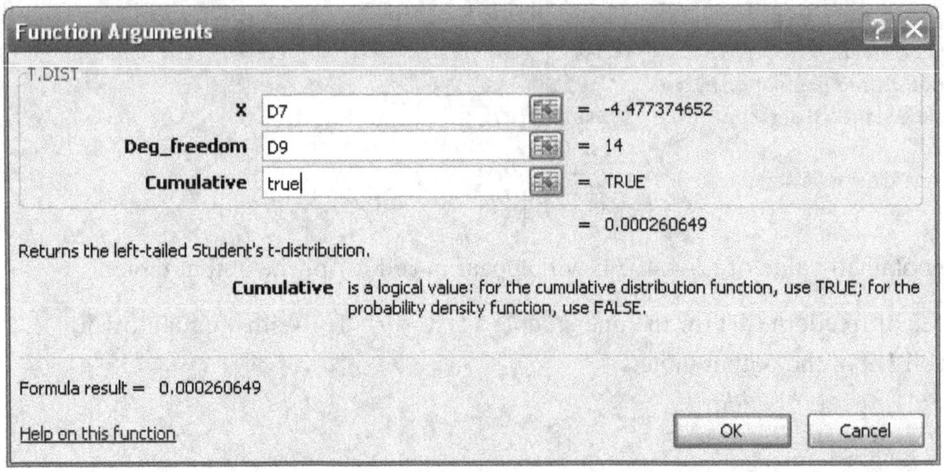

Function Arguments	? X
T.DIST	
X D7 = -4.477374652	
Deg_freedom D9 = 14	
Cumulative true = TRUE	
= 0.000260649	
Returns the left-tailed Student's t-distribution.	
Cumulative is a logical value: for the cumulative distribution function, use TRUE; for the probability density function, use FALSE.	
Formula result = 0.000260649	
Help on this function OK Cancel	

- Be certain that the flashing I-beam is positioned in the **X** window and click in cell **D7**. Or if you prefer, you can manually enter **D7** (or **-4.4774**) in the window.

- Click in the **Deg_freedom** window. Then click in cell **D9** to place the value of 14 in the window. If you prefer, you can manually enter **D9** (or **14**) in the window.

- Click in the **Cumulative** window. Enter **true** for the cumulative distribution function.

22. Click **OK**. The one-tailed probability of .0003 should now be displayed in cell D10 of the output table.

	A	B	C	D
1	HOURS		ONE-SAMPLE T-TEST	
2	15		Sample mean	14.53333
3	7		Hypothesized population mean	28
4	32		Sample standard deviation	11.64883
5	26		Count	15
6	31		Standard error of the mean	3.007715
7	5		t	-4.47737
8	0		Alpha	0.01
9	14		df	14
10	11		Probability one-tailed	0.000261

23. You will find the one-tailed critical value of t by using the **T.INV** function. Activate cell **D11** in the output table. Click **Insert Function** in the Formulas ribbon.

T.INV is designed to return the left-tailed inverse of the t distribution.

24. In the Insert Function dialog box, select the **Statistical** category and the **T.INV** function. Click **OK**.

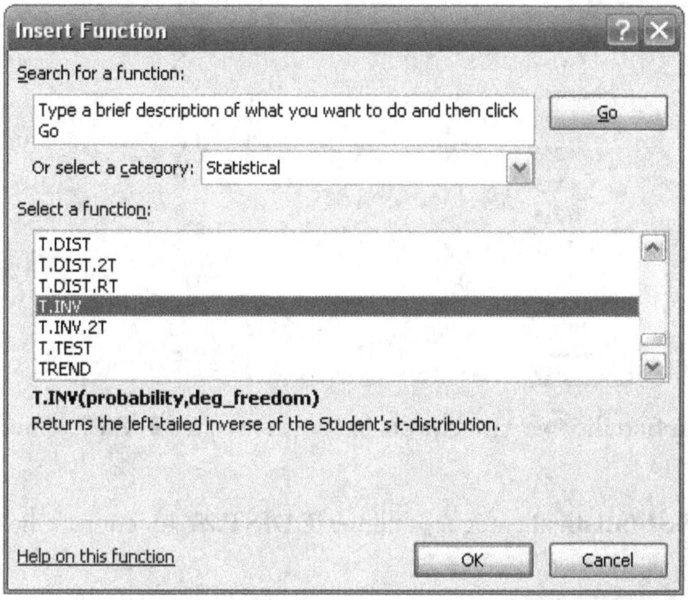

25. Click in the **Probability** window and click in cell **D8** in the output table to enter the cell address of alpha in the window. If you prefer, you can manually enter **D8** (or **0.01**) in the window.

26. Click in the **Deg_freedom** window and then click in cell **D9** in the output table to place the cell address of the df in the window. If you prefer, you can manually enter **D9** (or **14**) in the window.

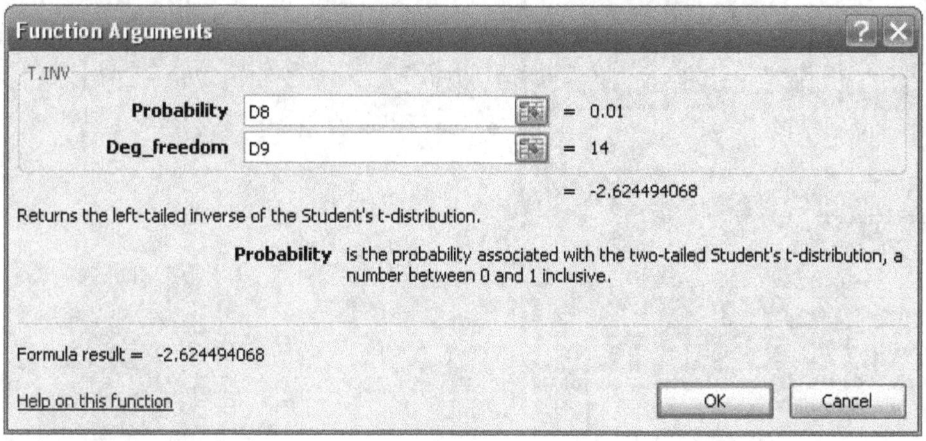

27.   Click **OK**. A critical one-tailed *t* of -2.6245 will now be displayed in cell D11 of the output table.

	A	B	C	D
1	HOURS		ONE-SAMPLE T-TEST	
2	15		Sample mean	14.53333
3	7		Hypothesized population mean	28
4	32		Sample standard deviation	11.64883
5	26		Count	15
6	31		Standard error of the mean	3.007715
7	5		t	-4.47737
8	0		Alpha	0.01
9	14		df	14
10	11		Probability one-tailed	0.000261
11	3		t critical one-tailed	-2.62449

28.   You will utilize the T.DIST.2T function to find the two-tailed probability. Activate cell **D12**. Click **Insert Function** in the Formulas ribbon.

29.   In the Insert Function dialog box, select the **Statistical** category and the **T.DIST.2T** function. Click **OK**.

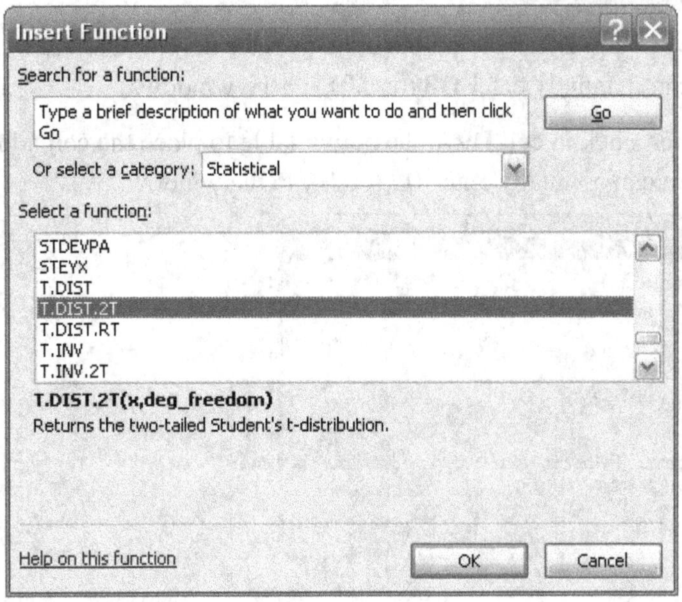

30. Complete the T.DIST.2T dialog box as shown below. An explanation of the entries is given immediately following the dialog box.

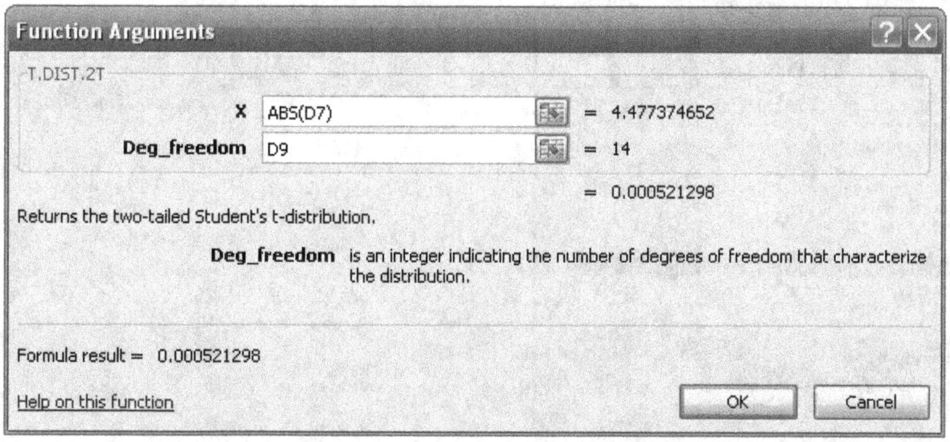

- Make sure that the flashing I-beam I located in the **X** window. Then enter **ABS(D7)**.

ABS is the absolute value function, and D7 is the address of the obtained t value. Note that you need the absolute value function here because the T.DIST.2T function is not designed to evaluate negative numbers.

- Click in the **Deg_freedom** window and then click in cell **D9**. If you prefer, you can manually enter **D9** or **14** in the window.

31. Click **OK**. A probability of .0005 is returned and placed in cell D12 of the output table.

	A	B	C	D
1	HOURS		ONE-SAMPLE T-TEST	
2	15		Sample mean	14.53333
3	7		Hypothesized population mean	28
4	32		Sample standard deviation	11.64883
5	26		Count	15
6	31		Standard error of the mean	3.007715
7	5		t	-4.47737
8	0		Alpha	0.01
9	14		df	14
10	11		Probability one-tailed	0.000261
11	3		t critical one-tailed	-2.62449
12	38		Probability two-tailed	0.000521

32. You will use the T.INV.2T function to obtain the two-tailed critical value of *t*. Activate cell **D13**. Click **Insert Function** in the Formulas ribbon.

33.   In the Insert Function dialog box, select the **Statistical** category and the **T.INV.2T** function. Click **OK**.

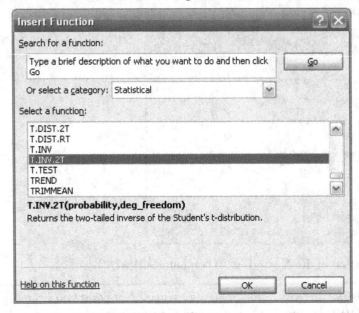

34.   Click in the **Probability** window and then click in cell **D8**, the cell containing alpha. Click in the **Deg_freedom** window and then click in cell **D9**, the cell containing df.

35.   Click **OK** and the two-tailed critical *t* of 2.9768 will appear in the output table.

Interpreting the Output

	A	B	C	D
1	HOURS		ONE-SAMPLE T-TEST	
2	15		Sample mean	14.53333
3	7		Hypothesized population mean	28
4	32		Sample standard deviation	11.64883
5	26		Count	15
6	31		Standard error of the mean	3.007715
7	5		t	-4.47737
8	0		Alpha	0.01
9	14		df	14
10	11		Probability one-tailed	0.000261
11	3		t critical one-tailed	-2.62449
12	38		Probability two-tailed	0.000521
13	10		t critical two-tailed	2.976843

- **Sample mean**. The mean hours of weekly TV viewing for the persons included in the sample is 14.5333.

- **Hypothesized population mean**. The value 28 indicates that the hypothesis under test in this analysis is that the population mean viewing time for adults residing in Canada is 28 hours.

- **Sample standard deviation**. The standard deviation of the sample is 11.6488. This value was computed using the formula for the unbiased estimate of the population standard deviation.

- **Count**. The number of observations in the sample is 15.

- **Standard error of the mean**. The standard error of the mean is the standard deviation of the sampling distribution of the mean. It provides an approximate average amount by which sample means differ from the population mean. For our research problem, we can say that the distribution of means for samples of $n = 15$ taken from the population of weekly TV viewing times of Canadian adults has a standard deviation of 3.0077.

- *t*. The value of t obtained by applying the formula shown at the beginning of this section is equal to -4.4774.

- **Alpha**. Alpha is the significance level for the statistical test. This value is selected by the researcher. Because we wanted to perform a relatively conservative test, we set alpha at .01 for this analysis.

- **df**. For this one-sample *t*-test, the formula for df is $n - 1$. With n equal to 15, df = 14.

- **Probability one-tailed**. This is the one-tailed chance probability of obtained t. More specifically, if the null hypothesis is true that the population mean TV viewing time is 28 hours per week, then the chance probability of obtaining a $t \leq -4.774$ is .0003. Similarly, for a value in the upper tail of the distribution, the chance probability of obtaining a $t \geq 4.774$ is .0003. Because the one-tailed probability of obtained t is less than alpha (.01), the one-tailed test would be considered statistically significant—assuming that the direction of the difference is consistent with the researcher's alternative hypothesis.

- *t* **critical one-tailed**. This is the absolute value of the one-tailed critical value of t for the selected value of alpha. With alpha equal to .01, and an alternative hypothesis that says that the population mean number of TV viewing hours for Canadian adults is less than 28, the one-tailed critical value of t is −2.6245.

Similarly, for an alternative hypothesis that says that the population mean for Canadian adults is greater than 28, the one-tailed critical value of t is 2.6245. If the alternative hypothesis stated that the mean for the Canadians was less than 28 hours, the one-tailed test would be declared statistically significant, because the absolute value of obtained t (4.4774) is greater than the absolute value of the one-tailed critical t.

- **Probability two-tailed**. This is the two-tailed chance probability associated with the obtained t. More specifically, if the null hypothesis is true that the population mean TV viewing time for Canadian adults is 28 hours, then the chance probability of obtaining a $t \leq -4.774$ plus the chance probability of obtaining a $t \geq 4.774$ is .0005. Note that this value is double that of the one-tailed probability. Because the two-tailed probability is less than alpha (.01), we would conclude that the two-tailed test is statistically significant.

- **t critical two-tailed**. This is the absolute value of the two-tailed critical value of t for the selected value of alpha. For alpha equal to .01, the two-tailed critical value of t is 2.9768. Since the absolute value of obtained t (4.4774) is greater than the absolute value of critical t, the two-tailed test result would be declared statistically significant.

Confidence Interval for the One-Sample t-Test

To compute the upper and lower limits of the confidence interval for the one-sample t-test, we use the following general expression:

$$\bar{X} \pm (t_{Crit})(S_{\bar{X}})$$

where $\bar{X}$ represents the sample mean, t_{Crit} represents the two-tailed t critical value, and $S_{\bar{X}}$ represents the standard error of the mean. To calculate the limits of the confidence interval, we will use the information provided in the ONE-SAMPLE T-TEST output table. We begin by setting up an output table to display a summary of our work.

1. For convenient reference, we'll place the CONFIDENCE INTERVAL output below the ONE-SAMPLE T-TEST output table. The table has labels for the lower limit and the upper limit.

12	38	Probability two-tailed	0.000521
13	10	t critical two-tailed	2.976843
14	9		
15	6	CONFIDENCE INTERVAL	
16	11	Lower Limit	
17		Upper Limit	

2. Let's start with the lower limit. We will use the following formula to calculate the lower limit of the confidence interval: **=D2-(D13*D6)**. Activate cell **D16** and key in **=D2-(D13*D6)**. Press **[Enter]** and the calculated value of the lower limit, 5.5798, will appear in the output table.

In this expression, D2 is the cell location of the sample mean, D13 is the cell location of the two-tailed t critical value, * indicates multiplication, and D6 is the cell location of the standard error of the mean.

15	6	CONFIDENCE INTERVAL	
16	11	Lower Limit	=D2-(D13*D6)

3. The formula you will enter for the upper limit of the confidence interval is **=D2+(D13*D6)**. Activate cell **D17** and key in **=D2+(D13*D6)**. Press **[Enter]** and the calculated value of the upper limit, 23.4868, will appear in the output table.

15	6	CONFIDENCE INTERVAL	
16	11	Lower Limit	5.579839
17		Upper Limit	=D2+(D13*D6)

Interpreting the Confidence Interval

15	6	CONFIDENCE INTERVAL	
16	11	Lower Limit	5.579839
17		Upper Limit	23.48683

The confidence interval for a population mean provides the range of values that, with a specified degree of confidence, includes the unknown population mean. When the statistical test result is significant, the sample mean is considered to be a good estimate of the population mean. Therefore, the sample mean is used when computing the upper and lower limits of the confidence interval. Because alpha was set at .01, a 99% confidence interval was constructed. The 99% confidence interval for this TV viewing research problem looks like: $5.5798 \le \mu \le 23.4868$.

Based on the limits of this interval, we would claim that we are 99% confident that the true population mean falls between 5.5798 and 23.4868 hours. Expressed as a confidence level, if this analysis were carried out repeatedly, sampling all possible samples of 15 Canadian adults, 99% of the confidence intervals would contain the true population mean. Only 1% of the intervals would not contain the true population mean. Because the hypothesized mean value of 28 is not included in the confidence interval, it is consistent with the previous conclusion based on the value of obtained *t*. Namely, the result is statistically significant and the true population mean is a value other than 28.

Chapter 8

Testing Hypotheses About the Difference Between Two Means

Researchers frequently want to investigate the effects of experimental treatments or the performance differences between existing groups such as males and females. The two broad categories of tests where observations are in two groups are: 1) independent, where data have been collected in a manner such that observations are not related to one another and 2) dependent (sometimes called paired or correlated), where either the same person is measured twice under different conditions or pairs of subjects are chosen so that they are similar to each other. In either case, the data are summarized in the form of sample means that can be compared using a *t*-test or a *z*-test. First I present the independent samples *t*-test, then the dependent samples *t*-test, and finally the *z*-test for two independent samples.

▶ Section 8.1 | *t*-Test for Two Independent Samples

The independent samples *t*-test that is most commonly used is the one that utilizes a pooled variance in the calculation of the standard error of the difference. To justify pooling the variances of two samples, one needs to assume that the variances of the two populations are equal. If this assumption cannot be met, then an alternative test is available that does not use a pooled variance. I first present the independent samples *t*-test that assumes equal population variances and then the alternative test that assumes unequal population variances.

Variances Are Not Known and Are Assumed to Be Equal

The formula for the independent samples *t*-test employing a pooled variance is

$$t = \frac{\left(\bar{X}_1 - \bar{X}_2\right) - \left(\mu_1 - \mu_2\right)}{S_{\bar{X}_1 - \bar{X}_2}}$$

where $\left(\bar{X}_1 - \bar{X}_2\right)$ is the difference between the two sample means, $\left(\mu_1 - \mu_2\right)$ is the hypothesized difference between the population means, and $S_{\bar{X}_1 - \bar{X}_2}$ is the standard error of the difference.

The standard error is calculated using a pooled variance estimate. The formula for the pooled variance is

$$S_{pooled}^2 = \frac{\left(n_1 - 1\right)S_1^2 + \left(n_2 - 1\right)S_2^2}{\left(n_1 - 1\right) + \left(n_2 - 1\right)}$$

where S_1^2 is the variance of sample 1, S_2^2 is the variance of sample 2, n_1 is the number of observations in sample 1, and n_2 is the number of observations in sample 2. As shown in this formula, the pooled variance estimate is the weighted average of the sample variances where each variance is weighted by its respective degrees of freedom.

The formula for the standard error of the difference, then, is given by

$$S_{\bar{X}_1 - \bar{X}_2} = \sqrt{\frac{S^2_{pooled}}{n_1} + \frac{S^2_{pooled}}{n_2}}$$

Assumptions Underlying the Independent Samples *t*-Test

The statistical assumptions underlying the independent samples *t*-test are

1. Observations are randomly sampled from population 1 and population 2.

2. The sample of observations from population 1 is independent of the sample of observations from population 2.

3. Observations are normally distributed in both population 1 and population 2.

4. The variances of population 1 and population 2 are unknown but are equal.

Sample Research Problem

A researcher wanted to find out if dreaming increased as a result of taking three milligrams of Melatonin before going to sleep each night. Nineteen people were randomly assigned to one of two treatment conditions: Melatonin ($n = 10$) and placebo ($n = 9$). The number of dreams recalled each morning were reported and then tallied over a one-week period.

Steps to Follow to Analyze the Sample Research Problem

1. Open worksheet "Ch8_Dreams" on the Web site, or enter the data in an Excel worksheet as shown below.

	A	B
	Melatonin	Placebo
1	Melatonin	Placebo
2	21	12
3	18	14
4	14	10
5	20	8
6	11	16
7	19	5
8	8	3
9	12	9
10	13	11
11	15	

2. Click on the **Data** tab near the top of the screen and select **Data Analysis**.

If Data Analysis does not appear as a choice in the Data ribbon, you will need to load the Microsoft Excel ToolPak add-in. Follow the procedure on page 9.

3. Three different t-tests can be selected. Select **t-Test: Two-Sample Assuming Equal Variances** and click **OK**.

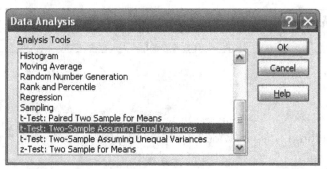

4. Complete the dialog box as shown below. A description of the entries is given immediately after the dialog box.

- **Variable 1 Range**. The flashing vertical line should be located in the **Variable 1 Range** window. Click in the top cell of the Melatonin group (cell **A1**) and drag to the end of the data in that column of the worksheet (cell **A11**). Or, if you prefer, manually enter **A1:A11** in the Variable 1 Range window.
- **Variable 2 Range**. Click in the **Variable 2 Range** window. Then click in cell **B1** and drag to the end of the data in that column (cell **B10**). If you prefer, you can manually enter **B1:B10** in the Variable 2 Range window.

You must click in the Variable 2 Range window before you drag over the range of scores, otherwise the cell addresses will be placed in the Variable 1 Range window.

- **Hypothesized Mean Difference**. This is the hypothesized difference between the population means. This difference is generally set equal to 0, although you could designate a nonzero value if you so desire. If no entry is made, the default value of zero is used. We'll use the default value.
- **Labels**. When checked, this indicates that the first cell in the ranges for sample 1 and sample 2 contains a variable label rather than a sample observation. In this example, these cells contain labels (Melatonin and Placebo). Therefore, click in the **Labels** box to place a check mark there.
- **Alpha**. This is the significance level for the statistical test. Let's use the default value of .05.
- **Output options**. Select the **Output Range option**. This will place the output in the same worksheet as the data. First click in the Output Range window. Then click in cell **A13** of the worksheet to place

the output a couple of lines below the data with **A13** as the uppermost left cell. If you prefer, you can manually enter **A13** in the window.

You could indicate the entire range in the Output Range window, but only the uppermost left cell is necessary.

---

5. Click **OK**.

Interpreting the Output

The output that you will obtain is shown below. If column A is not wide enough to display the output descriptions, you will want to increase the column width.

13	t-Test: Two-Sample Assuming Equal Variances	
14		
15		*Melatonin*	*Placebo*
16	Mean	15.1	9.777778
17	Variance	18.32222	16.94444
18	Observations	10	9
19	Pooled Variance	17.67386	
20	Hypothesized Mean Difference	0	
21	df	17	
22	t Stat	2.755318	
23	P(T<=t) one-tail	0.006758	
24	t Critical one-tail	1.739607	
25	P(T<=t) two-tail	0.013516	
26	t Critical two-tail	2.109816	

- **Mean**. Sample means for the two groups, Melatonin $(\overline{X}_1 = 15.1)$ and placebo $(\overline{X}_2 = 9.7778)$.

- **Variance**. Sample variances for group 1 $(S_1^2 = 18.3222)$ and group 2 $(S_2^2 = 16.9444)$, calculated using the formula for an unbiased estimate of the population variance.

- **Observations**. The number of observations in group 1 $(n_1 = 10)$ and in group 2 $(n_2 = 9)$.

- **Pooled Variance**. Pooled variance estimate $(S_{pooled}^2 = 17.6739)$, calculated by weighting each group variance by its degrees of freedom.

- **Hypothesized Mean Difference**. The value of 0 indicates that the null hypothesis states that the population means are equal (i.e., $\mu_1 - \mu_2 = 0$).

- **df**. Degrees of freedom for the test (df = 17), calculated by the formula $n_1 + n_2 - 2$.

- **t Stat**. Value of *t* obtained by applying the formula presented at the beginning of this section. For this research problem, obtained $t = 2.7553$.

- **P(T<=t) one-tail**. One-tailed chance probability of the obtained *t* statistic. This probability would be interpreted as follows. If the null hypothesis is true, then the chance probability of obtaining a $t \le -2.7553$ is .0068. Similarly, the chance probability of obtaining a $t \ge 2.5733$ is .0068. The result is statistically significant if the one-tailed probability is less than alpha.

- **t Critical one-tail**. One-tailed critical *t* value for selected alpha and df as indicated in the output. For this problem, the one-tailed critical value of *t* with df equal to 17 and alpha set at .05 is 1.7397. The result is statistically significant if the absolute value of obtained *t* is greater than critical *t*.

When you compare t-values to determine the statistical significance of a one-tailed test, you must check to be sure that the sign (positive or negative) of the obtained t value is consistent with your alternative hypothesis.

- **P(T<=t) two-tail**. Two-tailed chance probability of the obtained *t* statistic. If the null hypothesis is true, then the chance probability of obtaining a $t \leq -2.7553$ plus the chance probability of obtaining a $t \geq 2.7553$ is .0135. The result is statistically significant if the two-tailed probability is less than alpha.

- **t Critical two-tail**. Two-tailed critical *t* value for the selected alpha and df as indicated in the output. With df equal to 17 and alpha set at .05, the two-tailed critical value of *t* is 2.1098. The result is statistically significant if the absolute value of the obtained *t* is greater than critical *t*.

Variances Are Not Known and Are Assumed to Be Unequal

When the variances of the two populations are both unknown and assumed to be unequal, it may not be appropriate to use a pooled variance estimate. For the situation of unknown and unequal population variances, Excel employs the Welch-Aspin procedure, where the two sample variances, S_1^2 and S_2^2 are used in the calculation of the standard error of the difference. The obtained statistic is symbolized as t^*, for which degrees of freedom are not known but are approximated. The formula for t^* is given by

$$t^* = \frac{(\bar{X}_1 - \bar{X}_2) - (\mu_1 - \mu_2)}{S_{\bar{X}_1 - \bar{X}_2}}$$

which is the same as the test assuming equal variances, except for the calculation of the standard error of the difference. The standard error formula is

$$S_{\bar{X}_1 - \bar{X}_2} = \sqrt{\frac{S_1^2}{n_1} + \frac{S_2^2}{n_2}}$$

The degrees of freedom are approximated by the following formula:

$$df^* = \frac{df_1 df_2}{df_2 C^2 + df_1 (1 - C)^2}$$

The C term in this formula is found by

$$C = \frac{\left(\dfrac{S_1^2}{n_1}\right)}{\left(\dfrac{S_1^2}{n_1}\right) + \left(\dfrac{S_2^2}{n_2}\right)}$$

Assumptions Underlying the Independent Samples t^*-Test

The assumptions underlying the Welch-Aspin t^*-test are the same as those underlying the independent samples t-test, with the exception of the equality of population variances assumption. The Welch-Aspin t^* is based on the assumption that the population variances are not known and are not equal. In practice, the Welch-Aspin procedure is infrequently used, probably because of two considerations. First, when the sample sizes, n_1 and n_2, are equal, the pooled variances t-test is robust with respect to violations of the homogeneous variances assumption. Second, when sample sizes are both relatively large, say, greater than 30, the pooled variances t provides a satisfactory approximation. Thus, it is only when sample sizes are small and markedly unequal that t^* would be strongly recommended for analyzing data with unequal population variances.

Sample Research Problem

An animal researcher carried out an experiment designed to find out if the type of food reward given to rats for successfully negotiating a maze would have an effect on their maze-running times. Four rats were assigned to receive bran, an especially attractive food, and eight different rats were assigned to receive their regular food as a reward. The running times in seconds were recorded after 50 trials

Steps to Follow to Analyze the Sample Research Problem

1. Open the worksheet "Ch8_Rats" on the Web site, or enter the data in an Excel worksheet as shown below.

	A	B
1	Bran	Regular
2	32	118
3	38	78
4	27	82
5	45	91
6		67
7		41
8		97
9		42

2. Click on **Data Analysis** in the Data ribbon.

   If Data Analysis does not appear as a choice in the Data ribbon, you will need to load the Microsoft Excel ToolPak add-in. Follow the procedure on page 9.

3. In the Data Analysis dialog box, select **t-Test: Two-Sample Assuming Unequal Variances** and click **OK**.

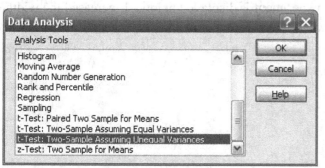

4. Complete the dialog box as shown below. Detailed instructions regarding the entries are given immediately following the dialog box.

- **Variable 1 Range**. Enter the worksheet location of the Bran condition. To do this, first click in the **Variable 1 Range** window. Then click in cell **A1** of the worksheet and drag to cell **A5**. If you prefer, you can manually enter **A1:A5**.
- **Variable 2 Range**. Enter the worksheet location of the Regular condition. To do this, first click in the **Variable 2 Range** window. Then click in cell **B1** of the worksheet and drag to cell **B9**. If you prefer, you can manually enter **B1:B9**.
- **Hypothesized Mean Difference**. We'll use the default value of zero. You do not have to enter anything here if you are using the default value.
- **Labels**. A check mark should appear in the box to the left of **Labels**, because the top cell in each of the variable ranges contains a label (Bran or Regular) rather than a data value.
- **Alpha**. Let's use the default value of .05 for the significance level of the statistical test.
- **Output options**. Let's place the output in a new worksheet. To do this, click the button to the left of **New Worksheet Ply** so that a black dot appears there.

5. Click **OK**.

Interpreting the Output

The output that you will obtain for the maze-running experiment is displayed below. I increased the width of column A so that I could read each label in its entirety.

	A	B	C
1	t-Test: Two-Sample Assuming Unequal Variances		
2			
3		Bran	Regular
4	Mean	35.5	77
5	Variance	60.33333	703.4286
6	Observations	4	8
7	Hypothesized Mean Difference	0	
8	df	9	
9	t Stat	-4.08888	
10	P(T<=t) one-tail	0.001361	
11	t Critical one-tail	1.833113	
12	P(T<=t) two-tail	0.002722	
13	t Critical two-tail	2.262157	

- **Mean**. Sample mean for the bran condition ($\overline{X}_1 = 35.5$) and sample mean for the regular condition ($\overline{X}_2 = 77$).

- **Variance**. Sample variance for the bran condition ($S_1^2 = 60.3333$) and the regular condition ($S_2^2 = 703.4286$).

- **Observations**. Number of observations in the bran condition ($n_1 = 4$) and the regular condition ($n_2 = 8$).

- **Hypothesized Mean Difference**. The hypothesized difference was set equal to 0, indicating that the hypothesis under test was that $\mu_1 - \mu_2 = 0$.

- **df**. Degrees of freedom for the t^*-test (df = 9), calculated via the Welch-Aspin procedure.

- **t Stat**. Value of t* obtained by applying the Welch-Aspin procedure. For this research problem, t* = – 4.0889.

- **P(T<=t) one-tail**. One-tailed chance probability of the obtained t*. If the null hypothesis is assumed to be true, then the chance probability of obtaining a $t^* \leq -4.0889$ is .0014. Similarly, .0014 is the chance probability associated with the obtained test statistic value in the upper tail of the distribution (i.e., $t^* \geq 4.0889$). A one-tailed test is statistically significant if the one-tailed probability is less than alpha.

- **t Critical one-tail**. One-tailed critical t^* value for designated alpha and df as indicated in the output. For this research problem, the one-tailed critical t^* value of 1.8331 is for alpha equal to .05 and 9 df. A one-tailed test is statistically significant if the absolute value of obtained t is greater than critical t.

When you compare t-values to determine the statistical significance of a one-tailed test, you must check to be sure that the sign (positive or negative) of the obtained t value is consistent with your alternative hypothesis.

- **P(T<=t) two-tail**. The two-tailed chance probability of obtained $t*$ equals .0027. It includes both the chance probability that $t* \leq -4.0889$ and $t* \geq 4.0889$, or double the one-tailed probability. A two-tailed test is statistically significant if the two-tailed probability is less than alpha.

- **t Critical two-tail**. Two-tailed critical $t*$ value for designated alpha and df as calculated according to the Welch-Aspin procedure. With df equal to 9 and alpha set at .05, the two-tailed critical value is 2.2622. A two-tailed test is statistically significant if the absolute value of obtained t is greater than critical t.

▶ *Section 8.2* | **Paired-Samples *t*-Test**

The paired-samples t-test is commonly used for two types of data analysis situations. The first is what is generally referred to as a before-after design, where the same research subjects are measured before and after an experimental treatment intervention. The second involves matching pairs of research subjects on a variable that is believed to be correlated with the dependent variable. For example, to analyze the data from an experiment on weight loss, persons could be matched on initial weight. Or, in a study on the effectiveness of instructional programs, students could be matched on intelligence test scores. The paired-samples *t*-test is probably more widely known to researchers as the correlated- or dependent-samples *t*-test. The problem that I use to illustrate the paired-samples *t* test utilizes the data collected in a study that employed a before-after design.

The formula for the matched-samples *t*-test is given by

$$t = \frac{\bar{D} - \mu_D}{S_{\bar{D}}}$$

where $\bar{D}$ is the sample mean difference score, μ_D is the hypothesized population mean difference, and $S_{\bar{D}}$ is the standard error of the mean difference.

The standard error of the mean difference is found by the following formula:

$$S_{\bar{D}} = \frac{S_D}{\sqrt{n}}$$

where S_D is the sample standard deviation of the difference scores, and n is the total number of pairs of observations.

Finally, the formula for S_D is given by

$$S_D = \sqrt{\frac{\sum(D - \bar{D})^2}{n-1}}$$

where D is a before-after difference score $(X_1 - X_2)$ and $\bar{D}$ is the mean difference score.

Assumptions Underlying the Paired-Samples *t*-Test

There are four assumptions underlying the paired samples t-test.

1. Observations are randomly sampled from the population(s) of interest.
2. The observations are correlated.
3. Observations are normally distributed in the population(s).
4. The variance of the difference scores is unknown.

Sample Research Problem

A researcher was concerned about the debilitating effects of test anxiety on academic performance and conducted a study to investigate the effectiveness of relaxation training for reducing anxiety. Six students who had been identified as having high test anxiety were randomly selected to participate in a relaxation training program. Before training began, each student completed a test developed to measure test anxiety. After the training was completed, each student again completed the same anxiety test. The scores obtained on the two administrations were recorded in an Excel worksheet

Steps to Follow to Analyze the Sample Research Problem

1. Open worksheet "Ch8_Anxiety" on the Web site, or enter the data in an Excel worksheet as shown below.

	A	B
1	Before	After
2	84	72
3	76	70
4	104	90
5	103	94
6	91	93
7	90	90

2. Click the **Data** tab near the top of the screen and select **Data Analysis**.

    If Data Analysis does not appear as a choice in the Data ribbon, you will need to load the Microsoft Excel ToolPak addin. Follow the procedure on page 9.

3. In the Data Analysis dialog box, select **t-Test: Paired Two Sample for Means** click **OK**.

    Data Analysis ? X

    Analysis Tools

    Histogram
    Moving Average
    Random Number Generation
    Rank and Percentile
    Regression
    Sampling
    t-Test: Paired Two Sample for Means
    t-Test: Two-Sample Assuming Equal Variances
    t-Test: Two-Sample Assuming Unequal Variances
    z-Test: Two Sample for Means

    OK
    Cancel
    Help

4. Complete the dialog box as shown here. Detailed information about the entries is given immediately following the dialog box.

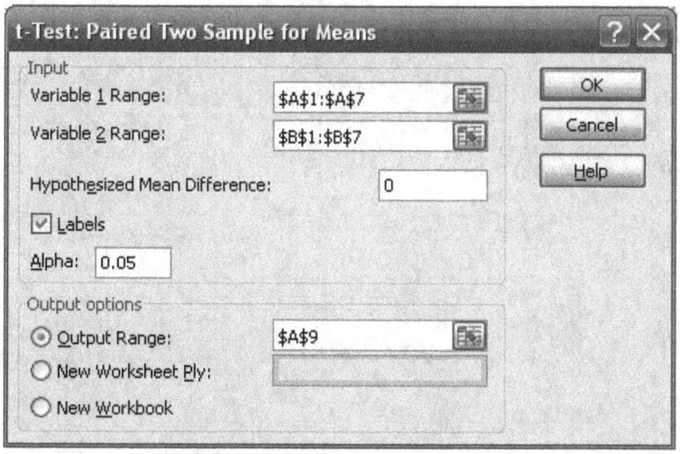

- **Variable 1 Range**. Enter the worksheet location of the Before condition. To do this, click in the **Variable 1 Range** window. Then, click in cell **A1** in the worksheet and drag to cell **A7**. If you prefer, you can manually enter **A1:A7**.
- **Variable 2 Range**. Enter the worksheet location of the After condition. Click in the **Variable 2 Range** window. Then click in cell **B1** of the worksheet and drag to cell **B7**. If you prefer, you can manually enter **B1:B7**.
- **Hypothesized Mean Difference**. Enter **0** to test the hypothesis that the population mean difference is equal to zero.
- **Labels**. The top cell in each of the variable ranges contains a label (Before and After), so a check mark should appear in the **Labels** box. Click in the box to place a check mark there.
- **Alpha**. Alpha refers to the significance level for the statistical test. Let's use the default value of .05.
- **Output options**. Let's place the output in the same worksheet as the data. Click in the **Output Range** window. Then click on cell **A9** in the worksheet. Cell A9 will be the uppermost left cell of the output. If you prefer, you can manually enter **A9** in the Output Range window.

5. Click **OK**.

Interpreting the Output

The output that you will obtain for the paired-samples *t*-test example will be similar to the output that is shown here.

	Before	After
9 t-Test: Paired Two Sample for Means		
10		
11	*Before*	*After*
12 Mean	91.33333	84.83333
13 Variance	117.4667	117.7667
14 Observations	6	6
15 Pearson Correlation	0.82358	
16 Hypothesized Mean Difference	0	
17 df	5	
18 t Stat	2.471525	
19 P(T<=t) one-tail	0.028212	
20 t Critical one-tail	2.015048	
21 P(T<=t) two-tail	0.056425	
22 t Critical two-tail	2.570582	

- **Mean**. Mean anxiety test score of the six students before they completed the relaxation training program ($\bar{X}_1 = 91.3333$) and the mean anxiety test score of the six students after they completed the program ($\bar{X}_2 = 84.8333$).

- **Variance**. Variance of the test taken before the relaxation training ($S_1^2 = 117.4667$) and variance of the test taken after training ($S_2^2 = 117.7667$), both calculated using the formula for an unbiased estimate of the population variance.

- **Observations**. Number of observations in the before group ($n_1 = 6$) and number of observations in the after group ($n_2 = 6$). The *n*'s will always be equal to one another for this *t*-test, regardless of whether you utilized a before-after design or a matched-pairs design.

- **Pearson Correlation**. The Pearson correlation coefficient, expressing the strength of the relationship between the pairs of before-after observations. The correlation of .8236 indicates a strong relationship between the pairs of observations.

- **Hypothesized Mean Difference**. This was designated as 0 in the dialog box, implying that the hypothesis under test was $\mu_1 - \mu_2 = 0$, or, alternatively, that the population mean difference (μ_D) was equal to 0.

- **df**. Degrees of freedom for the test (df = 5), calculated by the formula $n - 1$, where *n* refers to the total number of pairs of observations.

- **t Stat**. Value of *t* obtained by applying the formula for the paired-samples *t*-test. For the example analysis, obtained *t* = 2.4715.

- **P(T<=t) one-tail**. One-tailed chance probability of the obtained *t* statistic. More specifically, if the null hypothesis is true, then the chance probability of obtaining a $t \le -2.4715$ is .0282. Similarly, the chance probability of obtaining a $t \ge 2.4715$ is .0282. A one-tailed test is statistically significant if the one-tailed probability is less than alpha.

- **t Critical one-tail**. One-tailed critical *t* value for designated alpha (.05) and df equal to 5. For this analysis, the one-tailed critical *t* is equal to 2.0150. A one-tailed test is statistically significant if the absolute value of the obtained *t* is greater than critical *t*.

When you compare t-values to determine the statistical significance of a one-tailed test, you must check to be sure that the sign (positive or negative) of the obtained t value is consistent with your alternative hypothesis.

- **P(T<=t) two-tail**. Two-tailed chance probability of the obtained *t* statistic. That is, if the null hypothesis is true, then the chance probability of obtaining a $t \leq -2.4715$ plus the chance probability of obtaining a $t \geq 2.4715$ is .0564. A two-tailed test is statistically significant if the two-tailed probability is less than alpha.

- **t Critical two-tail**. Two-tailed critical value for designated alpha (.05) and df equal to 5. In this case, the critical two-tailed *t* is equal to 2.5706. A two-tailed test is statistically significant if obtained *t* is greater than the two-tailed critical *t*.

> ▶ Section 8.3   # *z*-Test for Two Independent Samples

The formula for the independent samples *z*-test is

$$z = \frac{(\bar{X}_1 - \bar{X}_2) - (\mu_1 - \mu_2)}{\sigma_{\bar{X}_1 - \bar{X}_2}}$$

where $(\bar{X}_1 - \bar{X}_2)$ is the difference between the two sample means, $(\mu_1 - \mu_2)$ is the hypothesized difference between the population means, and $\sigma_{\bar{X}_1 - \bar{X}_2}$ is the standard error of the difference.

The formula for the standard error of the difference is given by

$$\sigma_{\bar{X}_1 - \bar{X}_2} = \sqrt{\frac{\sigma_1^2}{n_1} + \frac{\sigma_2^2}{n_2}}$$

where σ_1^2 is the variance of population 1, σ_2^2 is the variance of population 2, n_1 is the size of sample 1, and n_2 is the size of sample 2.

Assumptions Underlying the *z*-Test

The assumptions underlying the two-samples *z*-test are similar to those for the independent samples *t*-test. An important difference, however, is that the *z*-test assumes that population variances are known. This assumption limits the usefulness of the *z*-test, because population parameters are seldom known and must be estimated from the sample data.

The statistical assumptions underlying the two-sample *z*-test include:

1. Observations are randomly sampled from population 1 and population 2.

2. The samples from the two populations are independent of each other.

3. Observations are normally distributed in population 1 and population 2.

4. The variances of the two populations, σ_1^2 and σ_2^2, are known.

Sample Research Problem

The mathematics performance of boys is generally believed to be higher than that of girls, especially at more advanced levels of mathematics. A researcher was interested in determining if this mathematics performance difference appeared as early as second grade. Nine second-grade boys and nine second-grade girls were randomly selected and administered a standardized mathematics achievement test for which it is known that the population variance for boys is 256 and the population variance for girls is 324.

Steps to Follow to Analyze the Sample Research Problem

1. Open worksheet "Ch8_Math" on the Web site, or enter the data in an Excel worksheet as shown here.

	A	B
1	Boys	Girls
2	45	51
3	34	36
4	51	24
5	46	43
6	51	37
7	31	39
8	48	52
9	25	37
10	50	48

2. Click on **Data Analysis** in the Data ribbon.

If Data Analysis does not appear as a choice in the Data ribbon, you will need to load the Microsoft Excel ToolPak add-in. Follow the procedure on page 9.

---

3. In the Data Analysis dialog box, select **z-Test: Two Sample for Means** and click **OK**.

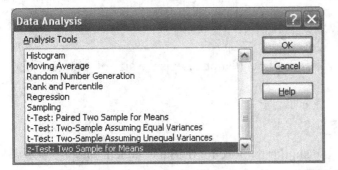

4. Complete the dialog box as shown here. A description of each entry is given immediately after the dialog box.

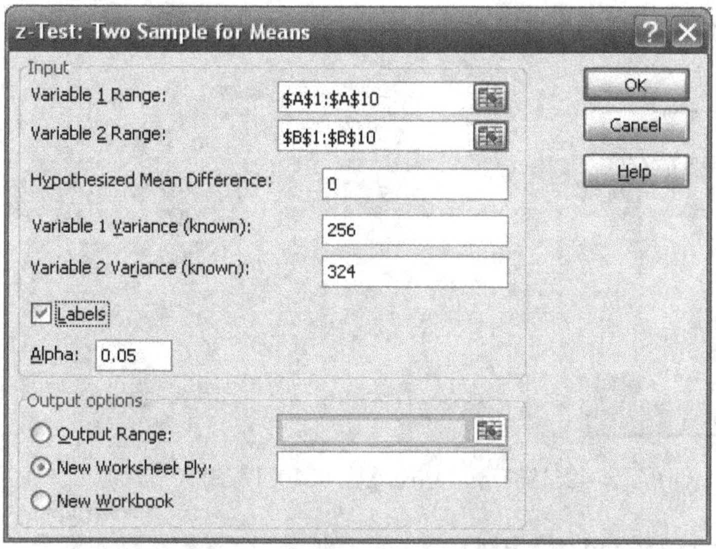

```
z-Test: Two Sample for Means                    [?] [X]
 Input
   Variable 1 Range:        [$A$1:$A$10    ] [▦]      [   OK   ]
   Variable 2 Range:        [$B$1:$B$10    ] [▦]      [ Cancel ]
   Hypothesized Mean Difference:  [0         ]         [  Help  ]
   Variable 1 Variance (known):   [256       ]
   Variable 2 Variance (known):   [324       ]
   [✓] Labels
   Alpha: [0.05 ]
 Output options
   ( ) Output Range:        [              ] [▦]
   (•) New Worksheet Ply:   [              ]
   ( ) New Workbook
```

- **Variable 1 Range**. The flashing vertical line should be located in the **Variable 1 Range** window. Click in the top cell of the Boys' data (cell **A1**) and drag to the end of the data in that column (**A10**). If you prefer, you can manually enter **A1:A10** in the Variable 1 Range window.
- **Variable 2 Range**. Click in the **Variable 2 Range** window. Then click in cell **B1** in the worksheet and drag to the end of the data in that column (**B10**). Or, if you prefer, you can manually enter **B1:B10** in the Variable 2 Range window.

You must click in the Variable 2 Range box before you drag over the range of scores, otherwise the addresses will be placed in the wrong location.

- **Hypothesized Mean Difference**. Enter a **0** for the value of the Hypothesized Mean Difference.
- **Variable 1 Variance (known)**. Enter **256** for the boys' population variance.
- **Variable 2 Variance (known)**. Enter **324** for the girls' population variance.
- **Labels**. Because group names (Boys and Girls) were included in the variable range, the **Labels** box must be checked. Click in the box to place a check mark there.
- **Alpha**. Let's use .05 for alpha, the default value.
- **Output options**. Let's place the output in a new worksheet. To do this, click in the button to the left of **New Worksheet Ply** so that a black dot appears there.
5. Click **OK**.

Interpreting the Output

You will receive output for this analysis that is similar to the output that is displayed here. If column A is not wide enough to display the output descriptions, you may want to increase the width.

	A	B	C
1	z-Test: Two Sample for Means		
2			
3		Boys	Girls
4	Mean	42.33333	40.77778
5	Known Variance	256	324
6	Observations	9	9
7	Hypothesized Mean Difference	0	
8	z	0.193773	
9	P(Z<=z) one-tail	0.423177	
10	z Critical one-tail	1.644854	
11	P(Z<=z) two-tail	0.846354	
12	z Critical two-tail	1.959964	

- **Mean.** Sample means for the two groups, boys $(\overline{X}_1 = 42.333)$ and girls $(\overline{X}_2 = 40.7778)$.

- **Known Variance.** Population variances, $\sigma_1^2 = 256$ and $\sigma_2^2 = 324$.

- **Observations.** Number of observations in group 1 ($n_1 = 9$) and in group 2 ($n_2 = 9$).

- **Hypothesized Mean Difference.** The value 0 indicates that the hypothesis under test in this analysis is that the population means of the second-grade boys and girls are equal.

- **z.** Value of z obtained by applying the formula presented at the beginning of this section. For this problem, obtained $z = 0.1938$.

- **P(Z<=z) one-tail.** One-tailed chance probability of the obtained z. That is, if the null hypothesis is true, then the chance probability of obtaining a $z \leq -0.1938$ is .4232. Or, in the upper tail of the distribution, the chance probability of obtaining a $z \geq 0.1938$ is .4232. The one-tailed test is statistically significant if the one-tailed probability is less than alpha.

- **z Critical one-tail.** One-tailed critical value of z for the alpha selected in the dialog box. In this case, with alpha set equal to .05, the one-tailed critical value of z is 1.6449. The one-tailed test is statistically significant if the absolute value of obtained z is greater than the one-tailed critical value.

When you compare z-values to determine the statistical significance of a one-tailed test, you must check to be sure that the sign (positive or negative) of the obtained z value is consistent with your alternative hypothesis.

- **P(Z<=z) two-tail.** Two-tailed chance probability of obtained z. If the null hypothesis is true, then the chance probability of obtaining a $z \leq -0.1938$ plus the chance probability of obtaining a $z \geq 0.1938$ is two times the one-tailed probability, $2 \times .4232 = .8464$. The two-tailed test is statistically significant if the two-tailed probability is less than alpha.

- **z Critical two-tail.** Two-tailed critical value of z for the alpha selected in the dialog box. With alpha set at .05, the two-tailed critical value is 1.9600. The two-tailed test is statistically significant if the absolute value of obtained z is greater than the two-tailed critical value.

Chapter 9
Analysis of Variance

Research designs often include more than one independent variable, as well as independent variables that have more than two levels. If a researcher is using such designs and is interested in differences among group means, analysis of variance (ANOVA) may be the appropriate statistical technique. Excel provides Data Analysis Tools for carrying out a one-way between-groups ANOVA, a one-way repeated measures ANOVA, and a two-way between-groups ANOVA.

► Section 9.1 | One-Way Between-Groups ANOVA

F-Test

To carry out a one-way between-groups ANOVA, one applies the F-test, which, for this analysis, is the ratio of mean square (MS) between to MS within:

$$F = \frac{MS_{Between}}{MS_{Within}}$$

Assumptions Underlying the F-Test

The statistical assumptions underlying the F-test include

1. Samples are randomly selected from the K populations.

2. The observations from the K populations are independent of one another.

3. Observations are normally distributed in the K populations.

4. The variances of the K populations are not known but are equal to one another.

Sample Research Problem

Sixteen adults were assigned to one of three conditions in which they were given a sheet with background information (e.g., sex, age, employment history) on a defendant who was being charged with child abuse. The three conditions were

1. **Clean record.** The information sheet stated that the defendant had a completely clean criminal record.

2. **Criminal record.** The information sheet stated that the defendant had a previous record of child abuse.

3. **Control.** The information sheet gave no information about the defendant's criminal record.

Adults in all three conditions were asked to view a video of the trial and then to rate their perception of the defendant's guilt or innocence. They were asked to utilize a 7-point rating scale, where 1 referred to "completely sure the defendant is guilty," and 7 referred to "completely sure the defendant is innocent." The researcher was interested in finding out if knowledge of a previous record would influence individuals' perceptions of guilt or innocence.

Using Analysis Tools for One-Way Between-Groups ANOVA

1.   Open worksheet "Ch9_Trial" on the Web site, or enter the participants' ratings in an Excel worksheet as shown below.

	A	B	C
1	Clean	Criminal	Control
2	6	1	4
3	5	2	3
4	7	1	5
5	6	3	6
6	3	2	4
7	7		

2.   Click the **Data** tab near the top of the screen and select **Data Analysis**.

If Data Analysis does not appear as a choice in the Data ribbon, you will need to load the Microsoft Excel ToolPak add-in. Follow the procedure on page 9.

3.   In the Data Analysis dialog box, select the Analysis Tool named **Anova: Single Factor**, and then click **OK**.

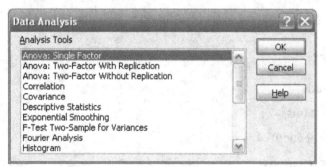

4.   Complete the Anova dialog box as shown below. Detailed information regarding the entries is given immediately after the dialog box.

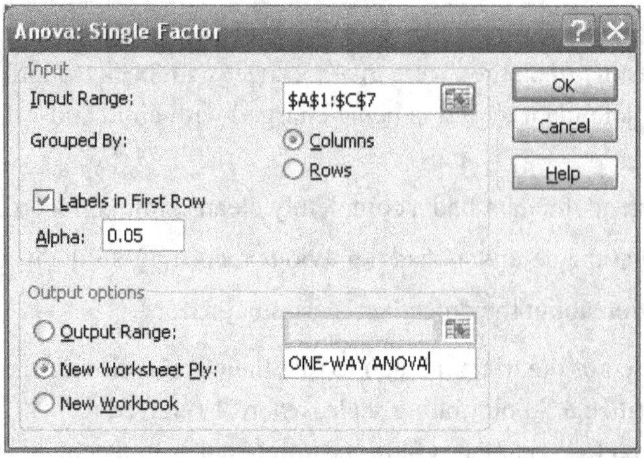

- **Input Range**. Enter the worksheet location of the data you wish to analyze. Click in the **Input Range** window. Then click in cell **A1** and drag to the end of the data in cell **C7**. If you prefer, you could also key in the data range by entering **A1:C7**. This input range includes the condition labels: Clean, Criminal, and Control.
- **Grouped by**. The data are grouped by columns in the worksheet. **Columns** was automatically selected, so you do not need to do anything additional for this step.
- **Labels in First Row**. The top cell in each column (A1, B1, and C1) of the data range contains a treatment condition label. To let Excel know that these are labels and should not be included in the data analysis, click in the box for **Labels in First Row.**
- **Alpha**. The default value of alpha is .05. If you want to carry out the test with a different alpha, just enter the desired significance level in the space to the right of **Alpha**. We will use the default value for this example.
- **Output options**. Three output options are available. Let's use the default option of **New Worksheet Ply**. The output table will be pasted into a new worksheet with A1 as the uppermost left cell. If you utilize the new worksheet option, you may give the worksheet a name that will enable you to identify it easily. Let's name the new worksheet **ONE-WAY ANOVA**.
5. Click **OK**. The output will be displayed in the ONE-WAY ANOVA worksheet.

	A	B	C	D	E	F	G
1	Anova: Single Factor						
2							
3	SUMMARY						
4	*Groups*	*Count*	*Sum*	*Average*	*Variance*		
5	Clean	6	34	5.666667	2.266667		
6	Criminal	5	9	1.8	0.7		
7	Control	5	22	4.4	1.3		
8							
9							
10	ANOVA						
11	*Source of Variation*	*SS*	*df*	*MS*	*F*	*P-value*	*F crit*
12	Between Groups	41.60417	2	20.80208	13.98761	0.000574	3.805565
13	Within Groups	19.33333	13	1.487179			
14							
15	Total	60.9375	15				

Interpreting the Output

The output provided for Anova: Single Factor (shown above) includes summary information about the groups and an ANOVA table.

SUMMARY

- **Groups**. The label for each of the three groups: Clean, Criminal, and Control. If the row with labels had not been included in the input range, the three groups would be identified as: Column 1, Column 2, and Column 3.

- **Count**. The number of observations in each group.

- **Sum**. The sum of the guilt/innocence ratings for each group.

- **Average**. The mean of the guilt/innocence ratings for each group.

- **Variance**. The variance of the guilt/innocence ratings for each group.

ANOVA Table

- **Source of Variation**. In a one-way between-groups ANOVA, the total variation is partitioned into two sources, between groups and within groups. Between groups refers to the degree to which the three groups differ from each other with respect to their guilt/innocence ratings. Within groups refers to the degree to which the adults within each group differ from one another.

- **SS**. Abbreviation for **sum of squares**.

    $SS_{Between}$ is obtained by applying the formula

    $$SS_{Between} = \sum_{K=1}^{K} \left(\overline{X}_K - \overline{X}_G \right)^2$$

    where $\overline{X}_K$ is the mean of the kth group, and

    $\overline{X}_G$ is the grand mean across all subjects.

    SS_{Within} is obtained by applying the formula

    $$SS_{Within} = \sum_{K=1}^{K} \left(n_K - 1 \right) S_K^2$$

    where n_K is the number of subjects in the kth group, and

    S_K^2 is the variance of the kth group.

- **df**. Abbreviation for degrees of freedom. $df_{Between} = K - 1$ and $df_{Within} = N - K$, where N refers to the total number of subjects.

- **MS**. Abbrevation for mean square. $MS_{Between}$ is computed by dividing $SS_{Between}$ by $df_{Between}$. Similarly, MS_{Within} is computed by dividing SS_{Within} by df_{Within}.

- **F**. The obtained value of the F test statistic, found by applying the formula

    $$F = \frac{MS_{Between}}{MS_{Within}}$$

    For the sample research problem,

    $$F = \frac{20.8021}{1.4872} = 13.9876$$

- **P-value**. The chance probability associated with the obtained value of F. If the null hypothesis is true that the population means for the three groups are all equal to one another, then the chance probability of obtaining an $F \geq 13.9876$ is .0006. Because the P-value is less than alpha (.05), we would conclude that the result is statistically significant and that the three population means are not all equal to one another.

- **F crit**. The critical value of F associated with the selected value of alpha (.05). Because the obtained F of 13.9876 is greater than the critical F of 3.8056, we would conclude that the result is statistically significant and that the three population means are not all equal to one another. Note that the conclusion reached by comparing the P-value to alpha should be the same conclusion that is reached by comparing the critical value of F to the obtained value of F.

► Section 9.2 One-Way Repeated Measures ANOVA

F-Test

The one-way between-groups ANOVA and the one-way repeated measures ANOVA differ with respect to how many times the subjects are measured or tested. Each subject is measured only once in the between-groups model. In the repeated measures model, each subject is measured three or more times. If subjects are measured just two times, the paired-samples *t*-test should be utilized. The *F*-test for the repeated measures model is given by

$$F = \frac{MS_{Treatments}}{MS_{Error}}$$

Assumptions Underlying the *F*-Test

1. The sample is randomly selected from the population of interest.
2. Each population of observations is normally distributed.
3. The variances of the K populations are not known but are equal to one another.
4. The population covariances for all pairs of treatment levels are equal.

Sample Research Problem

An investigator was interested in assessing the effects of drugs on driving skills. Six subjects performed the same driving simulation task under three different treatment conditions: 1) while under the influence of marijuana, 2) while under the influence of alcohol, and 3) while under the influence of no drugs. Driving simulation scores could range from 0 to 35, where higher scores were associated with better performance. The order of treatment administration was counterbalanced across the subjects. The data were recorded in an Excel worksheet with subject numbers in the first column and simulation scores in the next three columns.

Using Analysis Tools for One-Way Repeated Measures ANOVA

1. Open worksheet "Ch9_Drugs" on the Web site, or enter the data in an Excel worksheet as shown below.

	A	B	C	D
1	Subject	Marijuana	Alcohol	No Drug
2	1	19	8	21
3	2	18	10	31
4	3	25	10	26
5	4	20	18	28
6	5	17	7	14
7	6	21	16	24

2. Click the **Data** tab near the top of the screen and select **Data Analysis**.

If Data Analysis does not appear as a choice in the Data ribbon, you will need to load the Microsoft Excel ToolPak add-in. Follow the procedure on page 9.

3. In the Data Analysis dialog box, select **Anova: Two-Factor Without Replication** and click **OK**.

*"Anova: Two-Factor **Without Replication**" seems to be an unusual name for the one-way **repeated** measures ANOVA. The name, however, is appropriately descriptive. The two factors are subjects and treatments. "Without Replication" refers to n = 1 subject per cell. Note that this is also the procedure that would be utilized to perform a randomized blocks ANOVA, in which the two factors are blocks and treatments.*

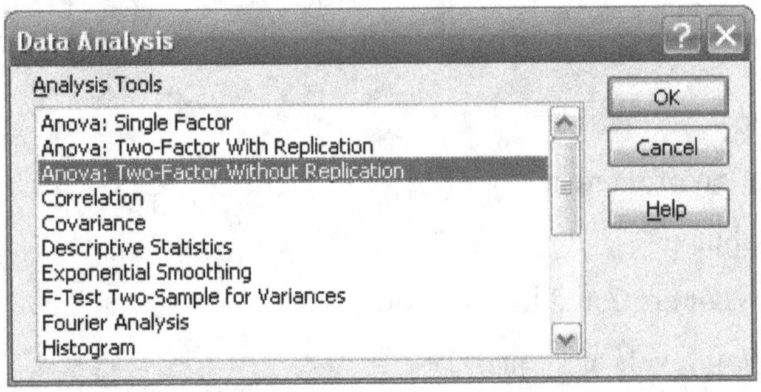

Data Analysis

Analysis Tools

Anova: Single Factor
Anova: Two-Factor With Replication
Anova: Two-Factor Without Replication
Correlation
Covariance
Descriptive Statistics
Exponential Smoothing
F-Test Two-Sample for Variances
Fourier Analysis
Histogram

OK   Cancel   Help

4. Complete the ANOVA dialog box as shown below. Detailed instructions are given immediately after the dialog box.

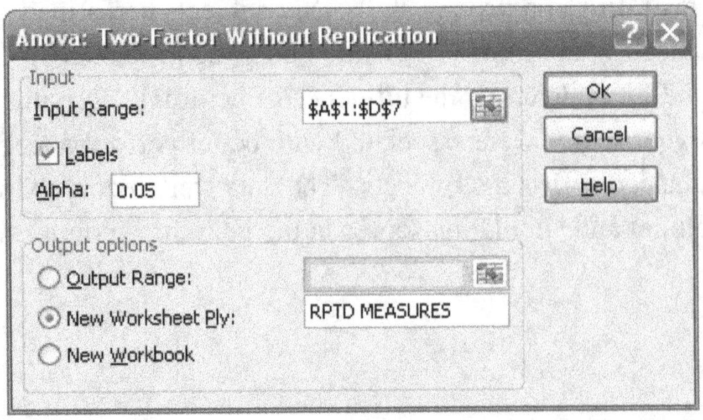

Anova: Two-Factor Without Replication

Input
Input Range: $A$1:$D$7

☑ Labels
Alpha: 0.05

Output options
○ Output Range:
◉ New Worksheet Ply: RPTD MEASURES
○ New Workbook

OK   Cancel   Help

- **Input Range**. Enter the worksheet location of the data you wish to analyze. Click in cell **A1** and drag to the end of the data in **D7**. You could also key in the data range by entering **A1:D7** or **$A$1:$D$7**.

It is essential that the first column contain identifying information for the subjects (ID numbers or names). If this column is not included in the data range, the output will be inaccurate.

- **Labels**. Click in the box to the left of **Labels** to indicate that the first cell in each of the three treatment columns is a label (i.e., Subject, Marijuana, Alcohol, No Drug) and not a value to be included in the analysis.
- **Alpha**. The default value for alpha is .05. Let's use the default value for this example.
- **Output options**. Select the default option, **New Worksheet Ply**. The output table will be pasted into a new worksheet with A1 as the uppermost left cell. If you select this option, you may give the worksheet an easily identifiable name. Let's name the worksheet **RPTD MEASURES**.
5. Click **OK**. The analysis output will be displayed in the RPTD MEASURES worksheet.

Interpreting the Output

The output provided for the one-way repeated measures ANOVA includes summary information about the subjects, summary information about the treatment conditions, and an ANOVA table. The output is shown below.

	A	B	C	D	E	F	G
1	Anova: Two-Factor Without Replication						
2							
3	SUMMARY	Count	Sum	Average	Variance		
4	1	3	48	16	49		
5	2	3	59	19.66667	112.3333		
6	3	3	61	20.33333	80.33333		
7	4	3	66	22	28		
8	5	3	38	12.66667	26.33333		
9	6	3	61	20.33333	16.33333		
10							
11	Marijuana	6	120	20	8		
12	Alcohol	6	69	11.5	19.9		
13	No Drug	6	144	24	35.6		
14							
15							
16	ANOVA						
17	Source of Variation	SS	df	MS	F	P-value	F crit
18	Rows	181.8333	5	36.36667	2.68059	0.086618	3.325835
19	Columns	489	2	244.5	18.02211	0.000483	4.102821
20	Error	135.6667	10	13.56667			
21							
22	Total	806.5	17				

SUMMARY

- **Count**. The total number of scores recorded for each subject, and, at the bottom of the summary section, the total number of scores recorded for each treatment condition. For this investigation, six subjects were tested three times each.

- **Sum**. The sum of each subject's scores across the three treatment conditions, and, at the bottom, the sum of the scores in each condition.

- **Average**. Each subject's mean score, and, at the bottom, the mean score obtained by the six subjects in each condition.

- **Variance**. The variance of each subject's scores, and, at the bottom, the variance of the six scores in each condition.

ANOVA Table

- **Source of Variation**

  Rows. Each subject's scores were recorded in a separate row of the worksheet, making subjects the row factor for this problem.

  Columns. Each treatment occupied a separate column in the worksheet, making treatments the column factor for this problem.

  Error. Error is the subjects-by-treatments interaction.

- **SS**

  Sum of squares total in the repeated measures model is partitioned into $SS_{Subjects}$, $SS_{Treatments}$, and $SS_{Interaction}$.

  For clarity in presentation of the formulas, I first computed I, II, III, IV, and V as shown below, and then utilized these values to compute each SS.

  $$I = \sum_{N=1}^{N} X_N^2 \text{, where } N \text{ refers to the total number of observations } (N = 18).$$

  $$I = 19^2 + 8^2 + 21^2 + \ldots + 24^2 = 6967$$

  $$II = \frac{\left(\sum_{N=1}^{N} X\right)^2}{N} \text{, where } N \text{ again refers to the total numbers of observations.}$$

  $$II = \frac{(333)^2}{18} = 6160.5$$

  $$III = \frac{\sum_{K=1}^{K} T^2}{n} \text{, where T is the sum of the scores in each treatment condition, K is the number of treatments,}$$

  and n is the number of subjects in each level of the treatment variable.

  $$III = \frac{(120)^2}{6} + \frac{(69)^2}{6} + \frac{(144)^2}{6} = 6649.5$$

  $$IV = \frac{\sum_{n=1}^{n} S^2}{K} \text{, where S is the sum of the scores for each subject.}$$

$$IV = \frac{(48)^2}{3} + \frac{(59)^2}{3} + \frac{(61)^2}{3} + ... + \frac{(61)^2}{3} = 6342.3333$$

$$SS_{Subjects} = IV - II = 181.8333$$

$$SS_{Treatments} = III - II = 489$$

$$SS_{Error} = I + II - III - IV = 135.6667$$

$$SS_{Total} = I - II = 806.5$$

- **df**. The formulas for df are as follows:

$df_{Subjects} = (n-1)$, where n is the number of subjects.

$$df_{Subjects} = (6-1) = 5$$

$df_{Treatments} = (K-1)$, where K is the number of treatments.

$$df_{Treatments} = (3-1) = 2$$

$$df_{Error} = (n-1)(K-1)$$

$$df_{Error} = (6-1)(3-1) = 10$$

$$df_{Total} = nK - 1$$

$$df_{Total} = (6)(3) - 1 = 17$$

- **MS**. The mean square terms are found by dividing each SS by its respective df.

- **F**. The obtained F associated with the subjects factor typically is not evaluated. The F associated with the treatments factor (columns) is found by dividing $MS_{Treatments}$ by MS_{Error}. Here, $F_{Treatments} = 18.0221$.

- **P-value**. Chance probability of obtained F if the null hypothesis is true that the three treatment means are equal to one another in the population. Because the P-value of .0005 is less than .05 (alpha), we would conclude that the result is statistically significant.

- **F crit**. The critical value of F distributed with $df_{Treatments} = 2$ and $df_{Error} = 10$, and alpha set at .05. Because the obtained F of 18.0221 exceeds the critical F of 4.1028, we would conclude that the result is statistically significant.

► Section 9.3 | Two-Way Between-Groups ANOVA

A two-way ANOVA allows a researcher simultaneously to analyze two factors and the interaction between the two factors. The procedure provided in Data Analysis Tools requires that both factors are between groups. That is to say, each subject can be included in only one group. Further, the cell n's must be equal.

F-Test

Three different F-tests are associated with the two-way between-groups ANOVA. The F for the main effect of factor A is given by

$$F_A = \frac{MS_A}{MS_{Within}}$$

The F for the main effect of factor B is given by

$$F_B = \frac{MS_B}{MS_{Within}}$$

Finally, the F for the A×B interaction is given by

$$F_{AXB} = \frac{MS_{AXB}}{MS_{Within}}$$

Assumptions Underlying the *F*-Test

The same assumptions apply to each of the three F-tests.

1. Subjects are randomly sampled from the populations of interest.

2. Observations are distributed normally in each population representing the combination of a level of A with a level of B.

3. Population variances are equal across all combinations of a level of A with a level of B.

4. Observations are independent, both within and across treatment combinations.

Sample Research Problem

An educational researcher was interested in the mathematics performance of low-achieving middle-school children. Although it has been found that performance benefits are associated with small-group work in mathematics, the researcher wanted to find out if the sex composition of the small groups made a difference. Of specific interest were dyads, in which each low-achieving student worked along with one high-achieving student. Would it be more beneficial for low achievers if they worked with high-achieving students whose sex was the same as their own? The researcher randomly selected 15 boys and 15 girls who were low achievers in mathematics and assigned them in equal numbers to one of three treatment conditions to work on a set of mathematics problems: 1) work with one high-achieving student of the same sex, 2) work with one high-achieving student of the opposite sex, or 3) work alone. After 30 minutes, the low-achieving students were given a test on problems similar to those they had worked on in their treatment condition.

Using Analysis Tools for Two-Way Between-Groups ANOVA

1. Open worksheet "Ch9_Dyads" on the Web site, or enter the data in an Excel worksheet as shown below. It is necessary to utilize labels for the three levels of the treatment variable (Same, Different, Alone) as well as for the two levels of the sex variable (Boy, Girl). In addition, the data need to be sorted by sex. As you can see in the worksheet below, the boys' test scores were entered first and then the girls'. Here, the label **Boy** appears only in the first row of the boys' data and the label **Girl** appears only in the first row of the girls' data. The procedure will also perform accurately if a group label appears in every row of column A.

▲	A	B	C	D
1	Sex	Same	Different	Alone
2	Boy	7	9	4
3		9	11	5
4		10	14	11
5		12	9	8
6		8	7	2
7	Girl	14	6	10
8		8	4	9
9		11	5	4
10		12	11	9
11		10	3	8

2. Click **Data Analysis** in the Data ribbon.

If Data Analysis does not appear as a choice in the Data ribbon, you will need to load the Microsoft Excel ToolPak add-in. Follow the procedure on page 9.

---

3. In the Data Analysis dialog box, select **Anova: Two-Factor With Replication** and click **OK**.

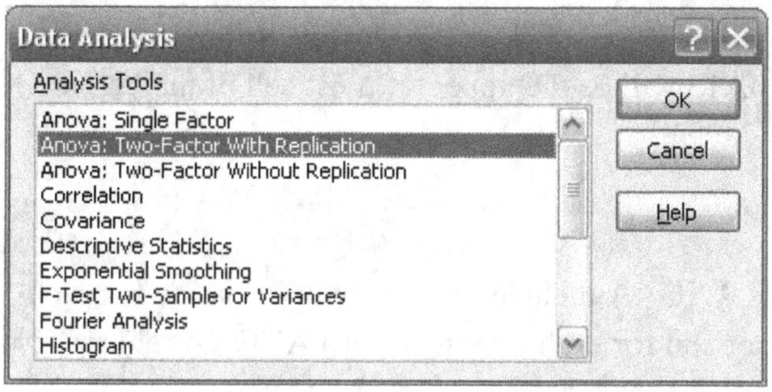

4. Complete the ANOVA dialog box as shown below. Detailed instructions are given immediately after the dialog box.

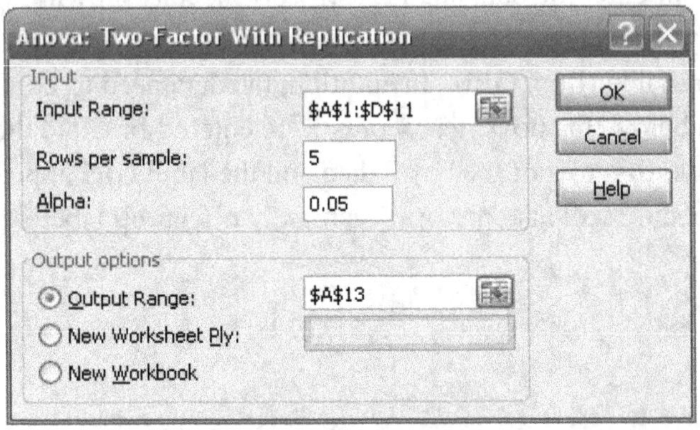

- **Input Range**. Enter the worksheet location of the data. Activate cell **A1** and drag to the end of the data in cell **D11**. You can also key in the data range by entering **A1:D11** or **$A$1:$D$11**.

Excel will not carry out this analysis procedure unless the range includes labels for the levels of the sex variable and labels for the levels of the treatment variable.

- **Rows per sample**. There must be the same number of rows in each sample. For this research problem, the five rows immediately below the treatment condition labels contain test scores for boys, and the next five rows contain test scores for girls.

Excel will not perform this analysis if cell n's are unequal or if the data set has any missing values.

- **Alpha**. Let's use the default value of .05.
- **Output options**. Let's place the output in the same worksheet with the data. Click in the button to the left of **Output Range** to place a black dot there. Next, click in the **Output Range** window. Then, click in cell **A13** of the worksheet. Cell A13 will be the upper leftmost cell of the output.
5. Click **OK**.

Interpreting the Output

The output provided for the two-factor between-groups ANOVA includes summary information for each gender-treatment combination as well as for each gender and for each treatment. An ANOVA table is also included.

SUMMARY

The SUMMARY portion of the output is displayed at the top of the next page.

13	Anova: Two-Factor With Replication				
14					
15	SUMMARY	Same	Different	Alone	Total
16	Boy				
17	Count	5	5	5	15
18	Sum	46	50	30	126
19	Average	9.2	10	6	8.4
20	Variance	3.7	7	12.5	9.828571
21					
22	Girl				
23	Count	5	5	5	15
24	Sum	55	29	40	124
25	Average	11	5.8	8	8.266667
26	Variance	5	9.7	5.5	10.6381
27					
28	Total				
29	Count	10	10	10	
30	Sum	101	79	70	
31	Average	10.1	7.9	7	
32	Variance	4.766667	12.32222	9.111111	

- **BOY**. The top portion of the summary section presents the count, sum, average, and variance for the boys in each of the three treatment conditions. The total column at the far right presents this same information for the boys across all three conditions. For example, the average performance for the boys in the Same, Different, and Alone conditions was 9.2, 10, and 6, respectively. The average score for boys across the three conditions was 8.4.

- **GIRL**. The next portion of the summary section presents the count, sum, average, and variance for the girls in each of the three treatment conditions. The total column at the far right displays this same summary information for the girls across the three conditions. For example, the variance of the girls' scores in the Same, Different, and Alone conditions was 5, 9.7, and 5.5, respectively. The variance of the scores of all 15 girls was 10.6381.

- **Total**. The last part of the summary section presents the count, sum, average, and variance for each treatment condition across sexes. For example, the sum of the scores in the Same, Different, and Alone conditions for the 10 boys and 10 girls combined was 101, 79, and 70, respectively.

ANOVA Table

The ANOVA table portion of the output is displayed below.

35	ANOVA						
36	Source of Variation	SS	df	MS	F	P-value	F crit
37	Sample	0.133333	1	0.133333	0.018433	0.893136	4.259677
38	Columns	50.86667	2	25.43333	3.516129	0.04579	3.402826
39	Interaction	62.06667	2	31.03333	4.290323	0.025528	3.402826
40	Within	173.6	24	7.233333			
41							
42	Total	286.6667	29				

- **Source of Variation**. In the two-way between-groups ANOVA, total variation is partitioned into sample variation (sex), column variation (treatment), interaction variation (sex × treatment interaction), and within groups variation.

- **SS**. Sex is the sample or row factor. I have designated sex as factor A in the sum of squares formulas. Treatment is the column factor and is designated as factor B. The sex × treatment interaction is referred to as A×B. I will first calculate I, II, III, IV, and V, and then use these values to compute each SS.

$$I = \sum_{N=1}^{N} X_N^2 \text{ , where } N \text{ refers to the total number of observations } (N = 30).$$

$$I = 7^2 + 9^2 + 4^2 + \ldots + 8^2 = 2370$$

$$II = \frac{\left(\sum\limits_{N=1}^{N} X\right)^2}{N} \text{ , where } N \text{ refers to the total number of observations.}$$

$$II = \frac{(250)^2}{30} = 2083.3333$$

$$III = \frac{\sum T_{A_i}^2}{n_i} \text{ , where } n_i \text{ is the number of observations in each level of factor A, and } T_{A_i} \text{ is the sum of the}$$

observations in each level of factor A.

$$III = \frac{126^2}{15} + \frac{124^2}{15} = 2083.4666$$

$$IV = \frac{\sum T_{B_j}^2}{n_j} \text{ , where } n_j \text{ is the number of observations in each level of factor B, and } T_{B_j} \text{ is the sum of the}$$

observations in each level of factor B.

$$IV = \frac{101^2}{10} + \frac{79^2}{10} + \frac{70^2}{10} = 2134.2$$

$$V = \frac{\sum T_{A_iB_j}^2}{n_{ij}} \text{ , where } n_{ij} \text{ is the number of observations in each cell, and } T_{A_iB_j} \text{ is the sum of the observations}$$

in each cell.

$$V = \frac{46^2}{5} + \frac{50^2}{5} + \frac{30^2}{5} + \ldots + \frac{40^2}{5} = 2196.4$$

$$SS_A = III - II = .1333$$

$$SS_B = IV - II = 50.8667$$

$$SS_{AXB} = II + V - III - IV = 62.0667$$

$$SS_{Within} = I - V = 173.6$$

$$SS_{Total} = I - II = 286.6667$$

- **df**. The formulas for df are as follows:

  $df_A = A - 1$, where A is the number of levels of factor A (sex).

  $df_A = 2 - 1 = 1$

  $df_B = B - 1$, where B is the number of levels of factor B (treatment).

  $df_B = 3 - 1 = 2$

  $df_{AXB} = (A - 1)(B - 1)$

  $df_{AXB} = (2 - 1)(3 - 1) = 2$

  $df_{Within} = AB(n_{ij} - 1)$, where n_{ij} is the number of observations in each cell.

  $df_{Within} = (2)(3)(5 - 1) = 24$

  $df_{Total} = N - 1$, where N is the total number of observations.

  $df_{Total} = 30 - 1 = 29$

- **MS**. The mean square terms are found by dividing each SS by its respective df.

- **F**. To obtain the three F-test values, MS_A, MS_B, and MS_{AXB} are each divided by MS_{Within}.

- **P-value**. Compare each P-value with .05 (alpha) to determine whether any of the three effects are statistically significant. For this problem, the F associated with factor B (treatment) and the F associated with the A×B interaction (sex × treatment) are significant because the P-values are both less than .05.

- **F crit**. If any of the three obtained F-test values exceed the critical F-test value, the result is statistically significant. For this research problem, the main effect of B (treatment) and the A×B interaction (sex × treatment) are both significant—the same conclusion that was reached by comparing P-values with alpha.

► Section 9.4 | *F*-Test for Two Sample Variances

The F-test is most often used in analysis of variance to test the significance of differences among group means. The F-test can also be used to test the hypothesis that two population variances are equal. The formula is

$$F = \frac{S_1^2}{S_2^2}$$

where S_1^2 refers to the variance of sample 1 and S_2^2 refers to the variance of sample 2. Numerator degrees of freedom are equal to $n_1 - 1$, and denominator degrees of freedom are equal to $n_2 - 1$. This F-test is carried out as a one-tailed test, assuming that you have specified in advance which of the two variances is larger. If you want to do a two-tailed test, you would place the larger of the two variances in the numerator and divide the desired alpha level in half. For example, if you wish to do a two-tailed test with alpha equal to .05, set the one-tailed alpha equal to .025.

Assumptions Underlying the *F*-Test

The assumptions underlying the F-test for two population variances are:

1. Samples are randomly selected from the two populations.

2. Observations are normally distributed in the two populations.

3. The data are measured on an interval or ratio scale.

Sample Research Problem

A few years ago, I asked the students in my undergraduate statistics classes to provide responses to a survey that asked questions about their "ideal selves," such as ideal weight and ideal number of friends. In one of the questions, I asked the students to estimate the square footage of their ideal home. After looking at some of the responses, I was fairly certain that men were much more accurate in estimating reasonable square footage than women. I decided to carry out an F test to find out if the variances of the men's and women's ideal square footages were equal.

Using Analysis Tools for the F-Test for Two Sample Variances

1. Open worksheet "Ch9_Ideal Home" on the Web site, or enter the data into an Excel worksheet as shown below.

	A	B
1	Men	Women
2	3400	40
3	2500	5000
4	1000	400
5	4000	100000
6	900	8000
7	1600	100
8	1800	25000
9	800	3600
10	2000	500

2. Click the **Data** tab near the top of the screen and select **Data Analysis**.

If Data Analysis does not appear as a choice in the Data ribbon, you will need to load the Microsoft Excel ToolPak add-in. Follow the procedure on page 9.

---

3. In the Data Analysis dialog box, select **F-Test Two-Sample for Variances** and click **OK**.

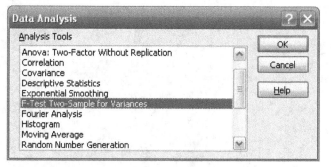

4. Complete the F-Test Two Sample for Variances dialog box as shown below. Detailed information about the entries is provided immediately after the dialog box.

- **Variable 1 Range**. When you are doing a two-tailed test, variable 1 should be the group with the larger variance. So, regardless of whether I am doing a two-tailed or one-tailed test, I usually designate the group with the larger variance as variable 1. For this example, the women's square footages have a larger variance than the men's. So, the women's square footages should be designated as variable 1. Click in the **Variable 1 Range** window. Then click in cell **B1** and drag to cell **B10**. If you prefer, you can manually enter **B1:B10**.
- **Variable 2 Range**. This is the range of the variable that has the smaller of the two group variances—the men's square footage responses. Click in the **Variable 2 Range** window. Then click in cell **A1** and drag to cell **A10**. If you prefer, you can manually enter **A1:A10**.
- **Labels**. The box to the left of **Labels** should be checked if the first row of the input contains labels that should not be included in the data analysis. The first row of our worksheet contains labels (Men and Women), so click in the box to place a check mark there.
- **Alpha**. Alpha refers to the type I error probability for the statistical test. Let's use the default value of .05.
- **Output options**. Let's use the default option of **New Worksheet Ply**. When we select this option, we can give the new worksheet a special name to make it easily identifiable. Let's use the name "Equal Variances Test." Click in the New Worksheet Ply window. Enter **Equal Variances Test**.

5. Click **OK**. The output that you will receive is displayed at the top of the next page.

Interpreting the Output

	A	B	C
1	F-Test Two-Sample for Variances		
2			
3		Women	Men
4	Mean	15848.89	2000
5	Variance	1.06E+09	1257500
6	Observations	9	9
7	df	8	8
8	F	841.6199	
9	P(F<=f) one-tail	6.92E-11	
10	F Critical one-tail	3.438101	

- **Mean**. The women's mean ideal square footage response is 15,848.89 and the men's is 2,000.

- **Variance**. The variance of the women's responses is 1.06E+09. Scientific notation is used because the variance is extremely large. If you move the decimal point 9 places to the right (i.e., +09), you see that the variance is 1,060,000,000. The variance of the men's responses is 1,257,500.

- **Observations**. The responses of 9 women and 9 men were included in the analysis.

- **df**. The degrees of freedom (df) for the numerator (women's responses) is equal to $n_1 - 1 = 9 - 1 = 8$. The df for the denominator (men's responses) is equal to $n_2 - 1 = 9 - 1 = 8$.

- **F**. The obtained F is equal to 841.6199.

- **P(F<=f) one-tail**. The chance one-tailed probability of obtained F is 6.92E–11. Scientific notation is used because the probability is extremely small. If you move the decimal point 11 places to the left, you see that the probability is .0000000000692. Because the probability of obtained F is smaller than alpha, the result is statistically significant.

- **F Critical one-tail**. The critical value of F for a one-tailed test is 3.4381. Because obtained F is greater than critical F, the result is statistically significant.

Correlation

Correlation coefficients are numerical indices that provide information regarding the relationship between two variables. In this chapter, I describe how to use features available in Excel for computing the Pearson correlation coefficient and the Spearman rank correlation. I also provide instructions for generating correlation matrices and creating scatterplots.

► Section 10.1 | **Pearson Correlation Coefficient**

The Pearson correlation coefficient, symbolized as r, ranges from -1, through 0, to $+1$. Coefficients close to -1 and $+1$ indicate strong linear relationships, whereas coefficients close to 0 indicate weak linear relationships. The correlation between two quantitative variables, X and Y, can be found by applying the formula

$$r_{XY} = \frac{Cov(X,Y)}{S_X S_Y}$$

where $Cov(X,Y)$ = the covariance of X and Y,

$\quad\quad S_X$ = the standard deviation of X, and

$\quad\quad S_Y$ = the standard deviation of Y.

Students of statistics may be more familiar with the computational formula, expressed as

$$r_{XY} = \frac{\sum XY - \frac{\left(\sum X\right)\left(\sum Y\right)}{N}}{\sqrt{\left(\sum X^2 - \frac{\left(\sum X\right)^2}{N}\right)\left(\sum Y^2 - \frac{\left(\sum Y\right)^2}{N}\right)}}$$

Sample Research Problem

An investigator was interested in parental influence on daughters regarding achievement in mathematics. Data were collected from 14 randomly selected intact families who had a 16-year-old daughter. Each daughter took a standardized mathematics achievement test. Each parent also completed a mathematics achievement test. The scores were entered in an Excel worksheet as displayed at the top of the next page.

	A	B	C
1	Daughter	Mother	Father
2	84	90	72
3	65	70	85
4	91	86	81
5	75	82	83
6	81	84	84
7	79	93	72
8	83	72	70
9	92	90	88
10	61	74	68
11	73	60	82
12	85	83	72
13	90	91	94
14	54	64	69
15	70	78	62

CORREL Function

Excel's CORREL function returns the value of Pearson's r for any two quantitative variables in a data set. No labels are attached to the r values, however, so I recommend that you enter appropriate labels in the worksheet.

1. Open worksheet "Ch10_Parents" on the Web site, or enter the data into an Excel worksheet as shown at the top of this page.

2. Let's find the correlation between daughter's achievement and mother's achievement. Type the label **Daughter-Mother** in cell A17.

Make column A wider so that the entire label will be displayed.

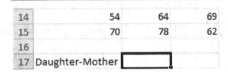

3. Activate cell **B17** where the r value will be placed.

4. Click the **Formulas** tab near the top of the screen and select **Insert Function**.

5. In the Insert Function dialog box, select the **Statistical** category, select the **CORREL** function, and click **OK**.

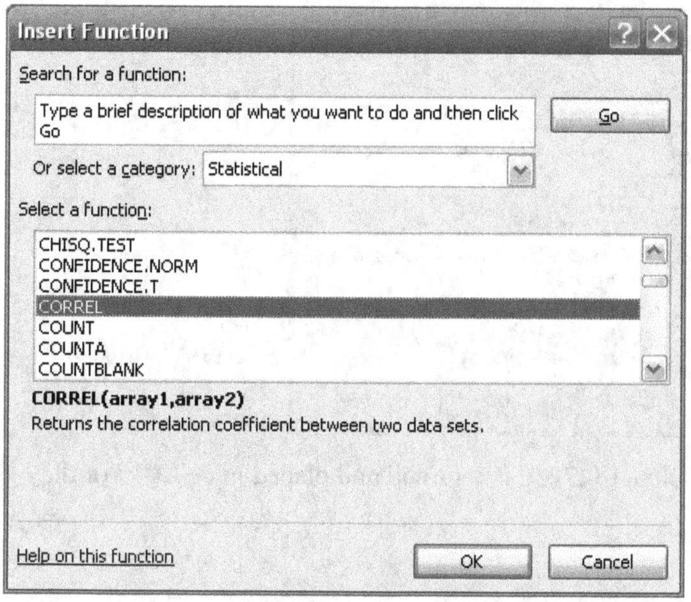

Insert Function   ? ✕

Search for a function:

| Type a brief description of what you want to do and then click Go | Go |

Or select a category: Statistical ▾

Select a function:

```
CHISQ.TEST
CONFIDENCE.NORM
CONFIDENCE.T
CORREL
COUNT
COUNTA
COUNTBLANK
```

CORREL(array1,array2)
Returns the correlation coefficient between two data sets.

Help on this function   OK   Cancel

6. The flashing I-beam should be positioned in the **Array 1** window. In the worksheet, click and drag over the range for Daughter, **A1** to **A15**. If you prefer, you can enter the range manually, **A1:A15**.

For the CORREL function, inclusion of the column label, cell A1 in this example, is optional.

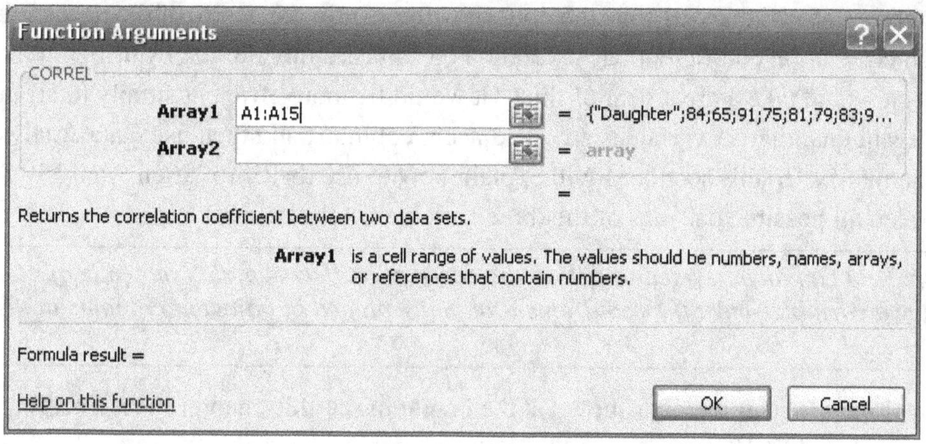

Function Arguments   ? ✕

CORREL

| **Array1** | A1:A15 | 📇 | = {"Daughter";84;65;91;75;81;79;83;9... |
| **Array2** | | 📇 | = array |

=

Returns the correlation coefficient between two data sets.

Array1 is a cell range of values. The values should be numbers, names, arrays, or references that contain numbers.

Formula result =

Help on this function   OK   Cancel

7. Click in the **Array 2** window. In the worksheet, click and drag over the range for Mother, **B1** to **B15**. Or, if you prefer, you can manually enter **B1:B15**.

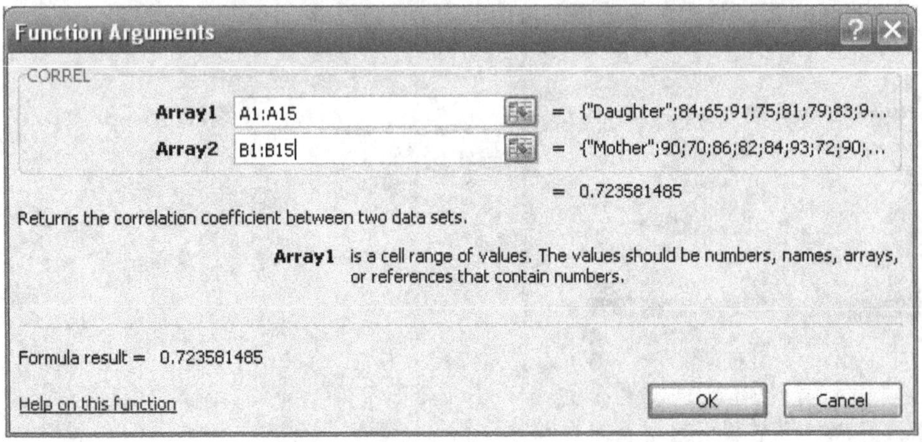

8. Click **OK** and a Pearson correlation coefficient of .7236 is returned and placed in cell B17 of the worksheet.

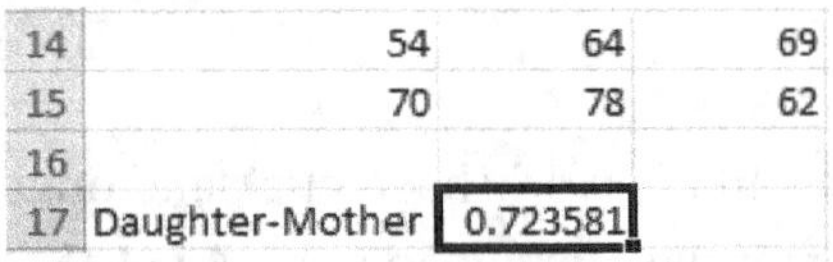

14		54	64	69
15		70	78	62
16				
17	Daughter-Mother	0.723581		

Correlation Matrices

When a data set includes a number of variables, as this one does, it is frequently the desire of the researcher to calculate not just one Pearson correlation coefficient but several. You can certainly do that by utilizing the CORREL function; however, it would take a great deal of time. It would be more efficient simply to arrange the worksheet so that the relevant quantitative variables are in adjacent columns, and then use Data Analysis Tools to produce a correlation matrix. In this section, I will explain how to use the Correlation Analysis Tool to compute correlations between all possible pairings of the three variables in the data set.

The Correlation Analysis Tool can also be used to obtain the correlation between only one pair of variables as long as the two variables meet the requirement of being placed in adjacent columns in the worksheet.

1. If you have not already done so, enter the data shown at the beginning of this chapter in an Excel worksheet. Or open worksheet "Ch10_Parents" on the Web site.

2. Click the **Data** tab near the top of the screen and select **Data Analysis**.

If Data Analysis does not appear as a choice in the Data ribbon, you will need to load the Microsoft Excel ToolPak add-in. Follow the procedure on page 9.

3. In the Data Analysis dialog box, select **Correlation** and click **OK**.

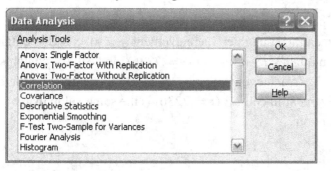

4. Complete the Correlation dialog box as shown below. A description of the entries is given immediately after the dialog box.

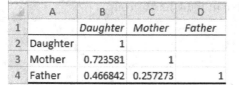

- **Input Range**. Click in the **Input Range** window. Then click on cell **A1** in the worksheet and drag to cell **C15**. If you prefer, you can manually enter **A1:C15**. Note that this range includes the entire data set which has three variables.
- **Grouped by**. The data in the worksheet are grouped by columns. Since **Columns** was already selected, you do not need to do anything additional here.
- **Labels in First Row**. Click in the box next to **Labels in First Row**. The check mark in this box lets Excel know that the first row contains variable labels (e.g., Daughter) that should not be included in the analysis.
- **Output options**. Select **New Worksheet Ply**. This is the default option. Unless specified otherwise, the correlation matrix will be pasted into a new worksheet with A1 as the upper left cell. If you select this option, you may give the worksheet a name that will enable you to identify it easily. Let's name the worksheet **Correlation**.

5. Click **OK**. The correlation matrix is generated and placed in the sheet named Correlation.

	A	B	C	D
1		*Daughter*	*Mother*	*Father*
2	Daughter	1		
3	Mother	0.723581	1	
4	Father	0.466842	0.257273	1

Interpreting the Output

Each entry in the matrix is the correlation between a pair of variables in the data set. For example, cell B2 contains the correlation between daughter's achievement and daughter's achievement ($r = 1$), cell B3 contains the correlation between daughter's achievement and mother's achievement ($r = .7236$), and cell C4 contains the correlation between mother's achievement and father's achievement ($r = .2573$). Note that the correlation between daughter's achievement and mother's achievement in cell B3 ($r = .7236$) is the same value that was obtained previously when utilizing the CORREL function.

Scatterplot

Excel provides an efficient way to produce a scatterplot of two variables. The procedure will not work, however, unless the two variables are adjacent to each other in the worksheet. In this section, I will show you how to produce a scatterplot of daughter's mathematics achievement and mother's mathematics achievement.

If you wish to produce a scatterplot of two variables that are not in adjacent columns of the worksheet, I suggest that you copy the variables of interest into a new worksheet rather than rearranging the entire data set. Cut, paste, or delete as necessary so that the two variables of interest are placed in adjacent columns.

1. Go back to the worksheet that contains the data. That worksheet is most likely Sheet 1. Click on the **Parents** (or **Sheet 1**) tab at the bottom of the screen.

2. Click and drag over the range **A1:B15**.

	A	B
1	Daughter	Mother
2	84	90
3	65	70
4	91	86
5	75	82
6	81	84
7	79	93
8	83	72
9	92	90
10	61	74
11	73	60
12	85	83
13	90	91
14	54	64
15	70	78

3. Click the **Insert** tab near the top of the screen and select **Scatter** in the Charts group.

4. Select the leftmost diagram in the top row.

5. Let's move the chart to a new sheet. To do this, **right-click** in the blank space of the chart near a border and select **Move Chart** from the menu that appears.

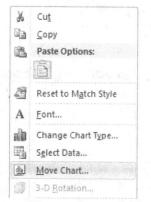

6. Select **New sheet** and click **OK.**

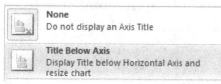

7. This scatterplot already has a chart title at the top (**Mother**) and a legend at the right. Let's change the chart title, remove the legend, and add a title to the Y-axis and the X-axis. Let's start with the chart title. Click on the word **Mother** so that it appears inside a frame.

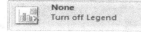

8. Delete **Mother** and replace it with **Mathematics Achievement of Daughters and Their Mothers**.

9. Click the **Layout** tab near the top of the screen and select **Legend** in the Labels group. Select **None** to turn off the legend.

10. Select **Axis Titles** in the Labels group of the Layout ribbon. Select **Primary Horizontal Axis Title** and **Title Below Axis**.

11. Delete **Axis title** and replace it with **Daughter's Score**.

12. Select **Axis Titles** in the Labels group of the Layout ribbon. Select **Primary Vertical Axis Title** and **Rotated Title**.

	None
	Do not display an Axis Title
	Rotated Title
	Display Rotated Axis Title and resize chart

13. Delete **Axis title** and replace it with **Mother's Score**.

The completed scatterplot is shown below.

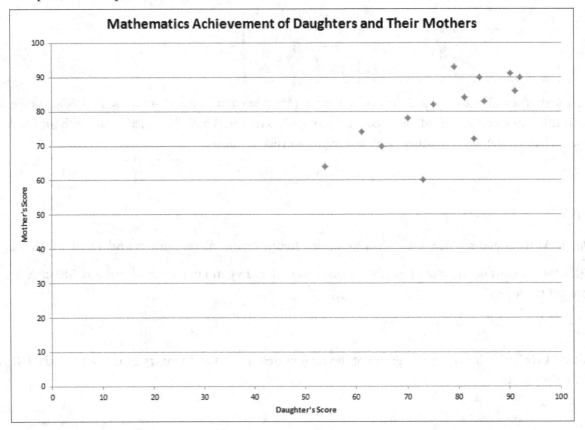

Modifying the Scatterplot

Although this scatterplot is accurate, it is unattractive, primarily because of the large amount of blank space to the left of the scatter of points. You can modify this as well as other aspects of the graph quite easily.

1. Let's begin by changing the minimum on the X-axis scale. The lowest achievement test score for the daughters in the data set is 54, so let's use 40 as the minimum. Move the pointer to the area of the graph that contains the X-axis scale (0 to 100). **Right-click** on any X-axis value. I right-clicked on 10. Select **Format Axis**.

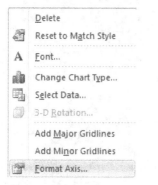

2. Select **Fixed** for the Minimum, and enter **40** in the window. Click **Close**.

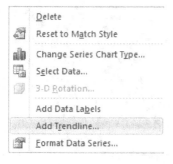

3. Let's also use 40 for the minimum value in the Y-axis scale. Move the pointer to the area of the graph that contains the Y-axis scale and **right-click** on any Y-axis value. Select **Format Axis**.

4. Select **Fixed** for the Minimum, and enter 40 in the window. Click **Close**.

5. Let's add vertical gridlines. Click the **Layout** tab near the top of the screen and select **Gridlines** in the Axes group.

6. Select **Primary Vertical Gridlines** and **Major Gridlines**. (The major horizontal gridlines should have been pre-selected.)

7. Next, we will add a trendline. **Right-click** on any dot in the scatterplot. Select **Add Trendline**.

8. **Linear** should be pre-selected.   At the bottom of the dialog box, we are given display options. Click the option to **Display R-squared value on chart**. Click **Close.**

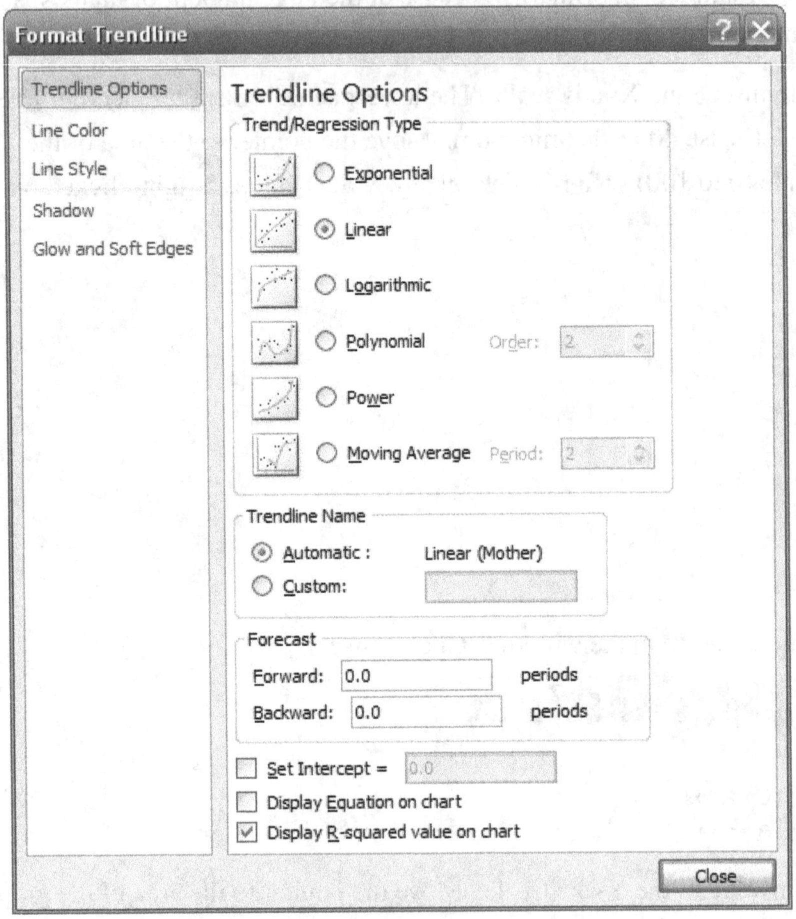

Format Trendline ? ✕

Trendline Options
Line Color
Line Style
Shadow
Glow and Soft Edges

Trendline Options

┌─ Trend/Regression Type ───────────────────┐

○ Exponential

● Linear

○ Logarithmic

○ Polynomial   Order: 2

○ Power

○ Moving Average   Period: 2

┌─ Trendline Name ──────────────────┐
● Automatic :   Linear (Mother)
○ Custom:

┌─ Forecast ──────────────────────┐
Forward: 0.0   periods
Backward: 0.0   periods

☐ Set Intercept =   0.0
☐ Display Equation on chart
☑ Display R-squared value on chart

Close

---

9. My R^2 value (.5236) was placed in the upper right hand corner and was difficult to read. I decided to move it to a different location. If you also would like to move the R^2 value, just click on it and drag it to a better location.   The completed scatterplot is shown at the top of the next page.   Several other modifications are possible. For example, you could change the font size, font style, or the appearance of the dots. Just right-click on the item that you would like to change and select from the shortcut menu that appears.

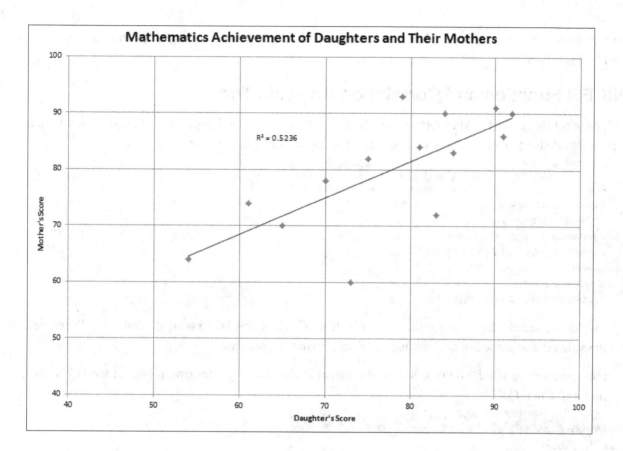

Spearman Rank Correlation Coefficient

Spearman's rank correlation coefficient is used when two variables are measured on an ordinal scale or two variables have been converted to ranks. The coefficient expresses the amount of agreement in the two sets of measures.

The formula for computing Spearman's correlation is

$$r_S = \frac{6\sum D^2}{N(N^2-1)}$$

where D = the difference in the ranks, and

N = the number of pairs of ranks.

Sample Research Problem

A researcher was interested in the extent to which men and women agreed on the quality of beers produced by microbreweries. The researcher prepared a list of seven microbrews and asked a sample of men and women to vote for the one that they considered the best overall. The researcher recorded the number of votes received by each beer.

RANK.EQ Function and Correlation Analysis Tool

1. Open worksheet "Ch10_Microbrews" on the Web site, or enter the labels and data shown below in an Excel worksheet. (Rank scores will be placed in the W Rank and M Rank columns.)

	A	B	C	D	E
1	Beer	Women	Men	W Rank	M Rank
2	Summit Extra Pale Ale	119	104		
3	Lift Bridge Farm Girl Saison	92	72		
4	Flat Earth Belgian Style Pale Ale	83	94		
5	Fulton Sweet Child of Vine	61	24		
6	Brau Brothers Sheephead Ale	31	10		
7	Schell Hopfenmalz	24	18		
8	Cold Spring Honey Almond Weiss	5	14		

2. First, let's generate the ranks for the women's data. Click in cell **D2** to activate that cell. Then click the **Formulas** tab near the top of the screen and select **Insert Function**.

3. In the Insert Function dialog box, select **Statistical** in the category window and select the **RANK.EQ** function. Click **OK**.

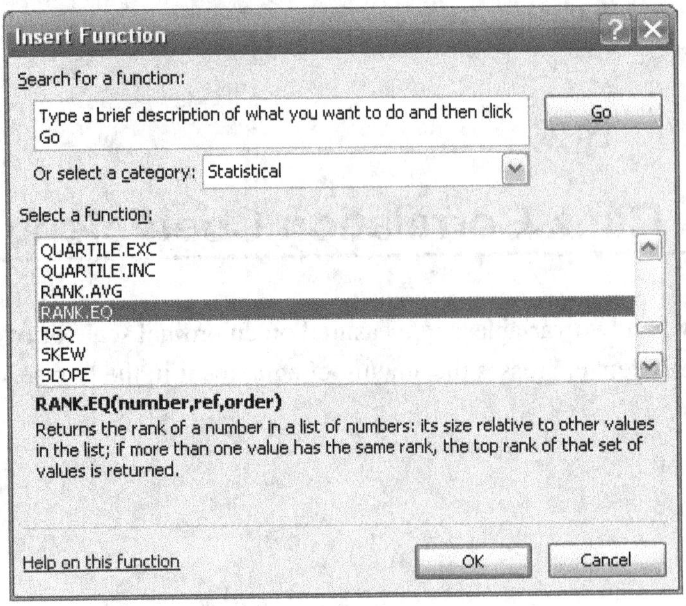

4. Click in the **Number** window and enter **B2**. Cell B2 is the worksheet address of the number of votes that Summit Extra Pale Ale received from women.

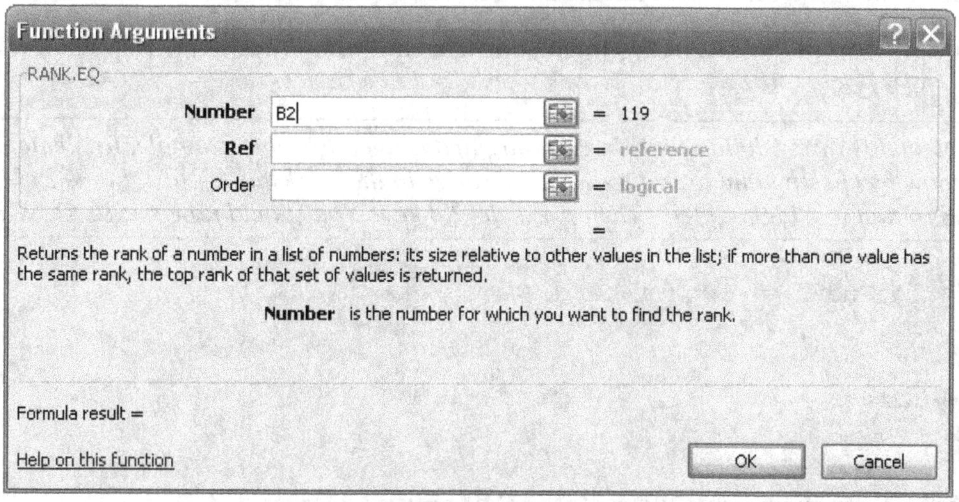

5. Click in the **Ref** window and enter **$B$2:$B$8**. What you are telling Excel to do is to find the rank of the number in cell B2 based on its position within the referenced range. Click **OK**.

It is very important that you include the dollar signs in the Ref window. The dollar signs make the range of cells absolute rather than relative. You want absolute cell references, because, in the next step, you will be copying the function in cell D2 to cells D3 through D8, and you want to use the same reference for all observations.

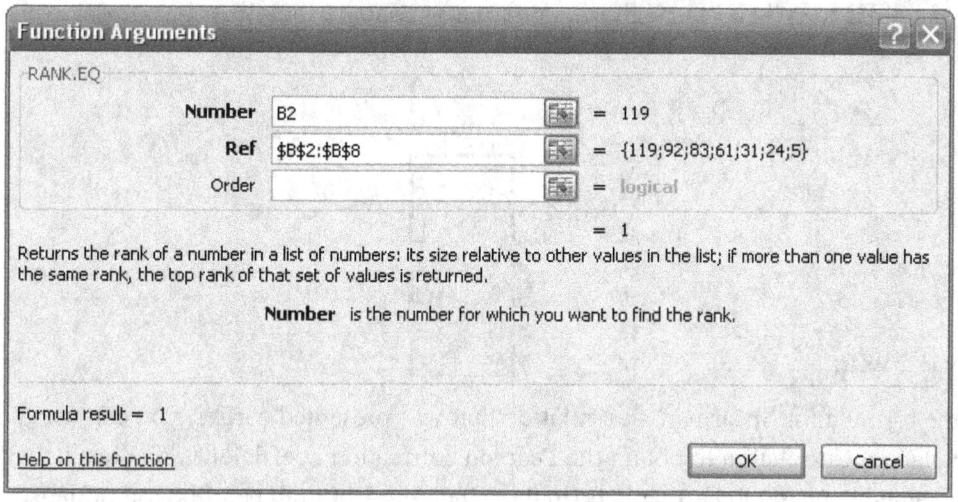

6. Copy the contents of cell D2 to cells D3 through D8.

	A	B	C	D	E
1	Beer	Women	Men	W Rank	M Rank
2	Summit Extra Pale Ale	119	104	1	
3	Lift Bridge Farm Girl Saison	92	72	2	
4	Flat Earth Belgian Style Pale Ale	83	94	3	
5	Fulton Sweet Child of Vine	61	24	4	
6	Brau Brothers Sheephead Ale	31	10	5	
7	Schell Hopfenmalz	24	18	6	
8	Cold Spring Honey Almond Weiss	5	14	7	

7. Click in cell **E2**. Then select **Insert Function** in the Formulas ribbon.

8. In the Insert Function dialog box, select **Statistical** and **RANK.EQ**. Click **OK**.

9. Click in the **Number** window of the RANK.EQ dialog box and enter **C2**. Click in the Ref window and enter **$C$2:$C$8**. Click **OK**.

The F4 key will automatically insert dollar signs in a highlighted range. For this example, first enter C2:C8 in the Ref window by clicking and dragging over the range in the worksheet. Next, in the Ref window, select C2:C8 so that it is highlighted. Then press the F4 key. You should now see $C$2:$C$8.

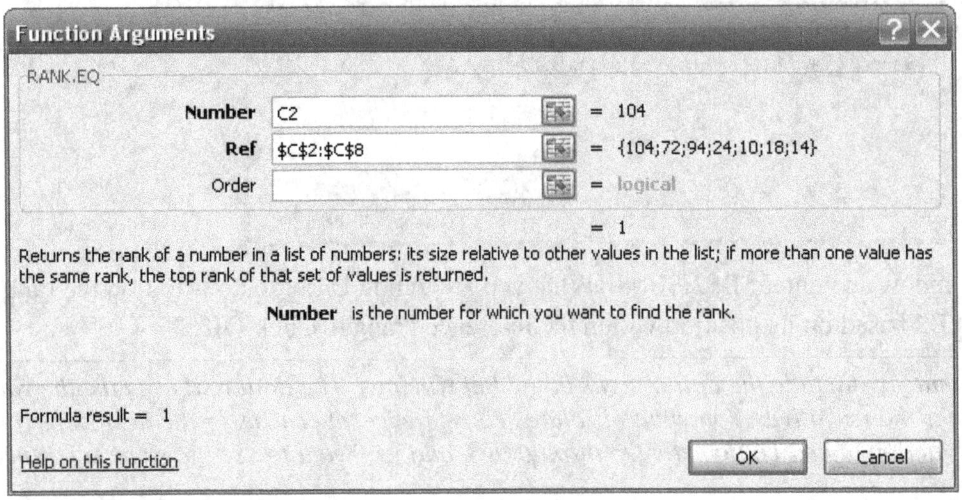

10. Copy the contents of cell E2 to cells E3 through E8. Your output should look like the output displayed below.

	A	B	C	D	E
1	Beer	Women	Men	W Rank	M Rank
2	Summit Extra Pale Ale	119	104	1	1
3	Lift Bridge Farm Girl Saison	92	72	2	3
4	Flat Earth Belgian Style Pale Ale	83	94	3	2
5	Fulton Sweet Child of Vine	61	24	4	4
6	Brau Brothers Sheephead Ale	31	10	5	7
7	Schell Hopfenmalz	24	18	6	5
8	Cold Spring Honey Almond Weiss	5	14	7	6

11. Rather than using the formula for Spearman's correlation that was presented earlier, you can use the Data Analysis Tool named Correlation to obtain the Pearson correlation coefficient. The results will be identical to those obtained by using Spearman's formula. Click the **Data** tab near the top of the screen and select **Data Analysis**.

If Data Analysis does not appear as a choice in the Data ribbon, you will need to load the Microsoft Excel ToolPak add-in. Follow the procedure on page 9.

---

12. In the Analysis Tools window, select **Correlation** and click **OK**.

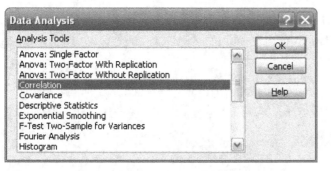

13. Complete the Correlation dialog box as shown below. Detailed information about the entries is provided immediately after the dialog box.

- **Input Range**. Enter the location of the ranks. Click in the **Input Range** window. Then activate cell **D1** and drag to the end of the rank data in cell **E8**. If you prefer, you can manually enter **D1:E8**.
- **Grouped by**. The data are grouped by columns in the worksheet. **Columns** is the default setting.
- **Labels in First Row**. A check mark must appear in this box to let Excel know that the top cell in each column of the input range contains a label and not a value that is to be included in the analysis.
- **Output options**. Select **Output Range** to place the output in the same worksheet as the data. Click in the Output Range window. Then enter cell **A10** in the window. A10 will be the uppermost left cell of the correlation output.

14. Click **OK**. The output you will receive is shown below.

10		W Rank	M Rank
11	W Rank	1	
12	M Rank	0.857143	1

Interpreting the Output

The Spearman rank correlation coefficient is .8571. This indicates fairly strong agreement between the women and men regarding the quality of the microbrews.

Chapter 11

Regression

► Section 11.1 | Two-Variable Regression

In a two-variable regression investigation, we have one predictor variable (X) that is used to predict a second variable (Y). The variable that is being predicted (Y) is referred to as the criterion variable. The prediction equation (also known as the regression line equation) is given by

$$\hat{Y} = a + bX$$

where $\hat{Y}$ = the predicted value of Y,

a = the Y-axis intercept,

b = the slope of the regression line, and

X = a value of the predictor variable.

The formula for computation of the intercept is

$$a = \bar{Y} - b\bar{X}$$

where $\bar{Y}$ = the mean of Y, and

$\bar{X}$ = the mean of X.

The formula for computation of the slope is

$$b = r_{XY} \cdot \frac{S_Y}{S_X}$$

where r_{XY} = the correlation between X and Y,

S_Y = the standard deviation of Y, and

S_X = the standard deviation of X.

Sample Research Problem

A university was interested in finding out how well ACT scores predicted the first-semester GPA of its freshmen. The worksheet containing the data for 15 freshmen is shown at the top of the next page.

⁄	A	B	C
1	ID	ACT	GPA
2	1	24	3.25
3	2	21	2.87
4	3	18	2.66
5	4	22	3.33
6	5	22	2.87
7	6	22	3.21
8	7	18	2.76
9	8	28	3.91
10	9	29	3.55
11	10	18	2.55
12	11	20	2.44
13	12	24	3.22
14	13	25	3.22
15	14	24	3.44
16	15	21	3.01

Steps to Follow to Analyze the Sample Research Problem

1. Open worksheet "Ch11_GPA" on the Web site, or enter the data into an Excel worksheet as shown above.

2. Click the **Data** tab near the top of the screen and select **Data Analysis**.

> *If Data Analysis does not appear as a choice in the Data ribbon, you will need to load the Microsoft Excel ToolPak add-in. Follow the procedure on page 9.*

3. In the Data Analysis dialog box, select **Regression** and click **OK**.

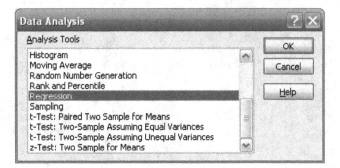

4. Complete the Regression dialog box as shown below. A description of the entries follows immediately after the dialog box.

- **Input Y Range** is the range in the worksheet containing scores on the criterion variable (GPA). To enter the range, first click in the **Input Y Range** window and then drag over cells **C1** through **C16**. If you prefer, the range can be entered manually by keying in **C1:C16**.

- **Input X Range** is the range for scores on the predictor variable (ACT). First click in the **Input X Range** window and then drag over the cells **B1** through **B16**. If you prefer, you can manually enter **B1:B16**.

- **Labels** should be checked because the variable names in cells B1 (ACT) and C1 (GPA) have been included in the input ranges and should not be included in the regression analysis. These labels, however, will be used in the output.

- **Constant is Zero** is not checked, because you do not want to force the regression line through the origin.

- **Confidence Level** is checked. Enter a value of **99** in the space to the right of Confidence Level. You will now see both 95% and 99% boundaries reported in the regression output. If Confidence Level is not checked, then only the default value of 95 will be utilized, and you will see the 95% boundaries reported twice in the output.

- **Output options** provide three choices. When you select an option, take into consideration that the regression output will take up at least seven columns.

    - **Output Range** is used when you want the regression output placed in the same worksheet as the data.

    - **New Worksheet Ply** will place the regression output in a new worksheet within the same book of sheets. If you are currently on sheet 1, this option will allow you to place the data on another sheet, with A1 as the upper left cell of the output.

    - **New Workbook** will place the regression output in a new workbook, with A1 as the upper left cell.

Click on Help in the Regression dialog box to obtain a description of the options and required entries.

---

5. Click **OK** and the output shown below will be generated. I recommend that you increase column width as necessary so that you can read the longer output labels.

	A	B	C	D	E	F	G	H	I
1	SUMMARY OUTPUT								
2									
3	*Regression Statistics*								
4	Multiple R	0.88358236							
5	R Square	0.780717786							
6	Adjusted R Square	0.763849924							
7	Standard Error	0.194760814							
8	Observations	15							
9									
10	ANOVA								
11		*df*	*SS*	*MS*	*F*	*Significance F*			
12	Regression	1	1.755646929	1.75565	46.2843	1.2562E-05			
13	Residual	13	0.493113071	0.03793					
14	Total	14	2.24876						
15									
16		*Coefficients*	*Standard Error*	*t Stat*	*P-value*	*Lower 95%*	*Upper 95%*	*Lower 95.0%*	*Upper 95.0%*
17	Intercept	0.72177665	0.3511329	2.05557	0.06049	-0.0367999	1.4803532	-0.0367999	1.48035316
18	ACT	0.105545685	0.015513989	6.80326	1.3E-05	0.07202975	0.1390616	0.07202975	0.13906162
22	RESIDUAL OUTPUT								
23									
24	*Observation*	*Predicted GPA*	*Residuals*						
25	1	3.254873096	-0.004873096						
26	2	2.938236041	-0.068236041						
27	3	2.621598985	0.038401015						
28	4	3.043781726	0.286218274						
29	5	3.043781726	-0.173781726						
30	6	3.043781726	0.166218274						
31	7	2.621598985	0.138401015						
32	8	3.677055838	0.232944162						
33	9	3.782601523	-0.232601523						
34	10	2.621598985	-0.071598985						
35	11	2.832690355	-0.392690355						
36	12	3.254873096	-0.034873096						
37	13	3.360418782	-0.140418782						
38	14	3.254873096	0.185126904						
39	15	2.938236041	0.071763959						

Interpreting the Output

SUMMARY OUTPUT

Regression Statistics

- **Multiple R**. Multiple R is the correlation between the predictor variable(s) and the criterion variable. Because we have only one predictor variable (ACT), .8836 is the Pearson correlation coefficient, r, expressing the linear relationship between ACT and GPA.

- **R Square**. *R* Square is also referred to as the coefficient of determination. It represents the proportion of variation in Y that is explained by its linear relationship with X. For our two-variable problem, R^2 is equal to .7807. Because we have only one predictor variable, R^2 could be represented as r^2.

- **Adjusted R Square**. The sample R^2 tends to be an optimistic estimate of the fit between the model and the population. Adjusted R^2 generally provides a better estimate. The adjusted R^2 for our sample two-variable problem is .7638.

- **Standard Error** is the standard error of estimate and is interpreted as the average error in predicting Y by means of the regression equation. For the sample problem, we would estimate that we err, on average, 0.1948 grade points when we use the regression equation to predict GPA from ACT.

- **Observations**. The number of observations refers to the number of subjects included in the analysis. Our analysis was carried out on the data provided by 15 subjects.

ANOVA

Regression analysis includes a test of the hypothesis that the slope of the regression line is equal to 0. If the slope is significantly different from 0, then we conclude that there is a statistically significant linear relationship between ACT and GPA.

- **Regression**. This component represents the variation in GPA that is explained by its relationship with ACT.

- **Residual**. Residual variation represents the variation in GPA that is not explained by ACT. It is considered "error variation" because it is unexplained by the predictor variable we have included in the analysis.

- **Total**. Total refers to "total variation." For this analysis, total variation is partitioned into regression variation and residual variation. Total variation, therefore, is the sum of regression variation and residual variation.

For each source of variation, the output provides degrees of freedom (df) and sums of squares (SS). The *F* value is obtained by dividing mean square (MS) regression by MS residual. The significance of *F* is the probability (P-value) associated with the obtained value of *F*. The test is statistically significant with alpha equal to .05 or .01, because the P-value of 1.25624E-05 is less than either of these values.

Coefficients

The information provided at the bottom of the output refers to the coefficients in the regression equation. The regression equation for the sample research problem is given by

$$\hat{Y} = 0.7218 + 0.1055X$$

If we utilize this formula to predict the freshman GPA for a student whose ACT score is 20, we would obtain

$$\hat{Y} = 0.7218 + 0.1055(20) = 2.83$$

- **Intercept**. The intercept is 0.7218. The **t Stat** refers to a test of the hypothesis that the intercept is significantly different from zero. The **P-value** is the probability associated with the obtained *t* statistic. The 95% and 99% confidence interval boundaries would be applied, respectively, to form a 95% and a 99% confidence interval around the intercept.

- **ACT**. The slope of the regression line is 0.1055. The **t Stat** refers to a test of the hypothesis that the slope is significantly different from zero. For the two-variable analysis, this t statistic and the F in the ANOVA table provide a test of the same hypothesis. Accordingly, F can be found by squaring the obtained t statistic. The **P-value** is the probability associated with the obtained t statistic. The 95% and 99% confidence interval boundaries can be used to form 95% and 99% confidence intervals, respectively, around the slope.

RESIDUAL OUTPUT

- **Observation**. The observations are numbered from 1 to 15. These numbers correspond to the order in which the observations appear in the data set. For example, observation 1 refers to subject 1 who received an ACT of 24 and a GPA of 3.25.

- **Predicted GPA**. The predicted GPA for each subject using the prediction equation $\hat{Y} = 0.7218 + 0.1055X$.

- **Residuals**. This is prediction error and is calculated $Y - \hat{Y}$. For example, the actual GPA, or Y, for observation 1 was 3.25. The predicted GPA, or $\hat{Y}$, for observation 1 was 3.2549. Prediction error for observation 1 was 3.25 – 3.2549, or –0.0049.

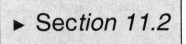 **► Section 11.2**

Multiple Regression

Multiple regression extends the concepts of two-variable regression to situations that include one criterion variable (Y) and two or more predictor variables, $X_1, X_2, X_3, \ldots, X_K$. Excel will allow you to include as many as 16 predictor variables. A model that includes two predictor variables is given by

$$\hat{Y} = B_0 + B_1 X_1 + B_2 X_2$$

Sample Research Problem

A personnel researcher wanted to find out how well sales volume of salespersons could be predicted from a score derived from demographic information (Biodata Score) and a Sales Aptitude test score. If the prediction equation worked well for individuals currently employed as salespersons, the equation could presumably be used for the selection of new employees. The data for 14 salespersons were entered in an Excel worksheet.

The predictor variables (Biodata Score and Sales Aptitude) must be entered in adjacent columns of the worksheet.

	A	B	C
1	Biodata Score	Sales Aptitude	Sales (000s)
2	25	35	234
3	33	44	124
4	44	65	345
5	54	55	467
6	45	60	432
7	23	33	235
8	53	54	555
9	22	12	146
10	42	34	345
11	25	23	366
12	16	22	156
13	23	16	123
14	36	55	564
15	33	28	237

Steps to Follow to Analyze the Sample Research Problem

1. Open worksheet "Ch11_Sales" on the Web site, or enter the data in an Excel worksheet as shown above.

2. Click the **Data** tab near the top of the screen and select **Data Analysis**.

If Data Analysis does not appear as a choice in the Data ribbon, you will need to load the Microsoft Excel ToolPak add-in. Follow the procedure on page 9.

3. In the Data Analysis dialog box, select **Regression** and click **OK**.

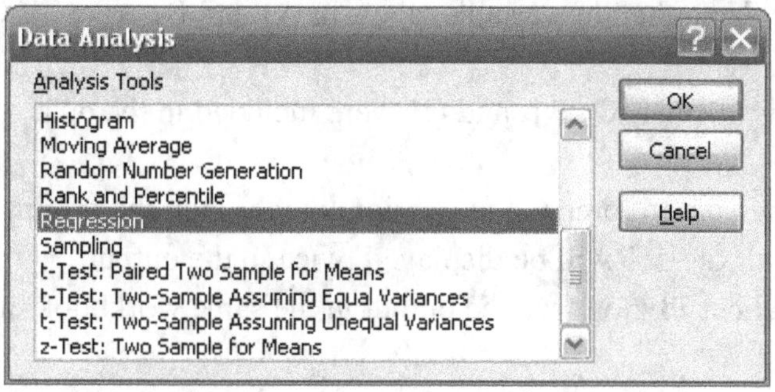

4. A completed Regression dialog box is shown below. A description of the entries follows immediately after the dialog box.

- **Input Y Range** is the range of the criterion variable, Sales(000s). First click in the **Input Y Range** window. Then click in cell **C1** in the worksheet and drag to cell **C15**. If you prefer, you can manually enter **C1:C15**. Note that the variable label in C1 was included in the input range.
- **Input X Range** includes scores on both predictor variables—Biodata Score and Sales Aptitude. Click in the **Input X Range** window. Then click in cell **A1** in the worksheet and drag to cell **B15**. If you prefer, you can manually enter **A1:B15**. Note that the variable labels in A1 and B1 were included in the input range.
- **Labels** is checked because the variable labels in cells A1, B1, and C1 were included in the data ranges.
- **Constant is Zero** is not checked. You do not want to force the regression line through the origin.
- **Confidence Level** is not checked. The default of 95% will be displayed twice in the output.
- Click in the button to the left of **New Worksheet Ply** to place the output in the same workbook but on a different sheet.
- **Residuals** are the differences between the actual data points and the values predicted by the regression equation. Click in the boxes to select all the residual output options.

Click on Help in the Regression dialog box to obtain a description of the options and required entries.

5. Click **OK**. The Summary Output provided by the regression procedure includes four sections: regression statistics, ANOVA summary table, coefficient summary information, and summary information regarding residuals. The lower 95% and upper 95% coefficient boundaries are repeated in the right-most sections of the output and are not included here.

	A	B	C	D	E	F	G
1	SUMMARY OUTPUT						
2							
3	*Regression Statistics*						
4	Multiple R	0.774351287					
5	R Square	0.599619916					
6	Adjusted R Square	0.526823537					
7	Standard Error	105.6322354					
8	Observations	14					
9							
10	ANOVA						
11		*df*	*SS*	*MS*	*F*	*Significance F*	
12	Regression	2	183818.4963	91909.25	8.236947	0.006510263	
13	Residual	11	122739.8608	11158.17			
14	Total	13	306558.3571				
15							
16		*Coefficients*	*Standard Error*	*t Stat*	*P-value*	*Lower 95%*	*Upper 95%*
17	Intercept	-11.3253401	86.32271854	-0.1312	0.897988	-201.3203626	178.6696824
18	Biodata Score	6.487621077	4.145268117	1.565067	0.145863	-2.636052534	15.61129469
19	Sales Aptitude	2.635116364	2.921508622	0.901971	0.386396	-3.795080758	9.065313487

Interpreting the Output

Regression Statistics

- **Multiple R**. The multiple correlation, R, between the two X variables (Biodata Score and Sales Aptitude) and Y [Sales(000s)] is .7744.

- **R Square**. R^2 is equal to .5996. Therefore, nearly 60 percent of the variation in Y is explainable by its linear relation with the two predictors.

- **Adjusted R Square**. As in the bivariate case, this is an adjusted value that more closely reflects the degree of fit in the population. For our sample problem, adjusted R^2 is equal to .5268.

- **Standard Error**. The standard error of estimate is equal to 105.6322. This is the average amount by which we err if we use the regression equation to predict Sales(000s).

- **Observations**. The data set includes 14 subjects with three variables recorded for each subject.

ANOVA

The ANOVA output is for a statistical test of the hypothesis of no linear relationship between the predictor variables and the criterion variable.

- **Regression**. The variation in Y that is explained by its relationship with the two predictor variables is referred to as *regression.*

- **Residual.** The variation in Y that is not explained by the two predictor variables is referred to as *residual.*

- **Total.** Total variation is partitioned into regression variation and residual variation.

MS regression and MS residual are found by dividing the appropriate SS by the respective df. The F value is obtained by dividing MS regression by MS residual. The significance of F refers to the probability (P-value) associated with obtained F. Statistical significance can be determined by comparing alpha with the P-value of .0065. The test would be considered significant with alpha equal to .05, as well as .01, because .0065 is less than either of these values.

Coefficient Summary Information

- **Intercept.** The intercept of the regression equation is –11.3253.

- **Biodata Score beta weight.** The beta weight for the Biodata Score is 6.4876.

- **Sales Aptitude beta weight.** The beta weight for Sales Aptitude is 2.6351.

When these values are placed in the regression equation, we see that the equation for predicting Sales(000s) from the two predictor variables is given by

$$\hat{Y} = -11.3253 + 6.4876X_1 + 2.6351X_2$$

If we use the data for the first person in the data set, we obtain

Predicted Sales (in 000s) $= -11.3253 + 6.4876(25) + 2.6351(35) = 243.0942$

- The **Standard Error** values, **t Stat**, **P-value**, and **Lower and Upper 95%** confidence interval boundaries are also provided for each of these coefficients. The t Stat for the intercept provides a statistical test of the hypothesis that the intercept is equal to zero. The t Stats for the predictor variable beta weights both test the hypothesis that the beta weight is equal to zero.

Residuals

The residual output is shown below.

23	RESIDUAL OUTPUT			
24				
25	Observation	Predicted Sales (000s)	Residuals	Standard Residuals
26	1	243.0942596	-9.094259552	-0.093593612
27	2	318.7112754	-194.7112754	-2.003871943
28	3	445.4125509	-100.4125509	-1.033396207
29	4	483.9375981	-16.93759806	-0.174313365
30	5	438.7245902	-6.724590194	-0.069206149
31	6	224.8487847	10.15121533	0.104471277
32	7	474.8148606	80.18513938	0.825225713
33	8	163.0237199	-17.02371994	-0.175199688
34	9	350.7487015	-5.748701492	-0.059162787
35	10	211.4728632	154.5271368	1.590316704
36	11	150.4491571	5.550842874	0.057126524
37	12	180.0518065	-57.05180648	-0.587148916
38	13	367.1604187	196.8395813	2.025775412
39	14	276.5494136	-39.54941362	-0.407022963

- **Predicted Sales(000s).** Predicted Sales is the value we will obtain when we utilize the prediction equation. Predicted Sales for the first person in the data set is 243.0943.

- **Residuals.** Residuals are found by subtracting the predicted value from the actual value. For the first person, $234 - 243.0943 = -9.0943$.

- **Standard Residuals.** Standard Residuals are found by transforming the residuals to a unit normal distribution with a mean of 0 and standard deviation of 1.

- **Residual Plots.** Residual plots were requested in the output options section of the Regression dialog box. These are shown below. These plots graphically compare actual Y with predicted Y for each of the two predictors. The charts can be selected, reshaped and printed on a full page if desired.

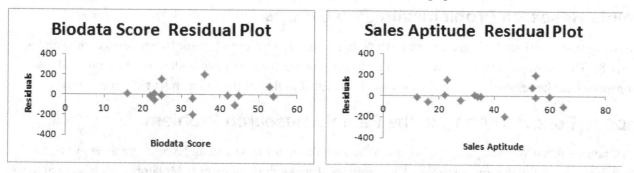

- **Line Fit Plots.** The optional line fit plots were also requested. When we have two predictors, two line fit plots are created. The actual and predicted values of Y are shown for the Biodata Scores in one of the plots and for Sales Aptitude Scores in the other plot.

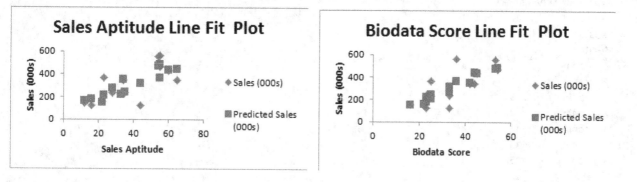

▶ Section 11.3	**Dummy Coding of Qualitative Variables**

When regression analysis is used for analysis of variance, the qualitative group membership variable is introduced into the regression equation by quantifying it. To quantify the group membership variable, first count the number of levels. If the number of levels is expressed as K, then the degrees of freedom associated with the group membership variable equal K – 1. Therefore, K – 1 linearly independent variables (X_1, X_2, X_3, ..., X_{K-1}) can be introduced into the regression model to represent the group membership variable. The general procedure is:

X_1 = 1 if the observation is a member of level 1, and X_1 = 0 if the observation is not a member of level 1.

X_2 = 1 if the observation is a member of level 2, and X_2 = 0 if the observation is not a member of level 2.

$X_3 = 1$ if the observation is a member of level 3, and $X_3 = 0$ if the observation is not a member of level 3, and so on up to X_{K-1}.

$X_{K-1} = 1$ if the observation is a member of level $K - 1$, and $X_{K-1} = 0$ if the observation is not a member of level $K - 1$.

Sample Research Problem with Two Groups

For a two-group problem, I will use the same data that were used for the *t*-test for two independent samples in Chapter 8. This example will not only illustrate dummy coding for a variable with two levels but will also demonstrate that the regression analysis outcome is identical to that of the independent samples *t*-test.

Steps to Follow to Analyze the Sample Research Problem

1. Open worksheet "Ch11_Dreams" on the Web site, or enter the data in an Excel worksheet as shown below. X1 is the dummy variable. X1 is equal to 1 when the treatment is Melatonin. X1 is equal to 0 when the treatment is not Melatonin.

	A	B	C
1	Treatment	Dreams	X1
2	Melatonin	21	1
3	Melatonin	18	1
4	Melatonin	14	1
5	Melatonin	20	1
6	Melatonin	11	1
7	Melatonin	19	1
8	Melatonin	8	1
9	Melatonin	12	1
10	Melatonin	13	1
11	Melatonin	15	1
12	Placebo	12	0
13	Placebo	14	0
14	Placebo	10	0
15	Placebo	8	0
16	Placebo	16	0
17	Placebo	5	0
18	Placebo	3	0
19	Placebo	9	0
20	Placebo	11	0

The X1 values can be keyed in directly or you can generate them by using an IF statement. In cell C2, enter =IF(A2="Melatonin",1,0). Press [Enter]. Then copy the contents of cell C2 to cells C3 through C20. The IF statement tells Excel that the C2 value should be "1" if A2 contains "Melatonin," and if A2 does not contain "Melatonin," then the C2 value should be 0.

2. Click the **Data** tab near the top of the screen and select **Data Analysis**.

If Data Analysis does not appear as a choice in the Data ribbon, you will need to load the Microsoft Excel ToolPak add-in. Follow the procedure on page 9.

3. In the Analysis Tools window, select **Regression** and click **OK**

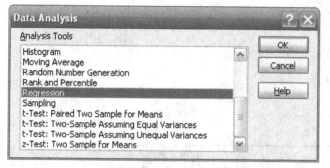

4. Complete the Regression dialog box as shown below. Detailed instructions are given immediately following the dialog box.

- **Input Y Range**. Enter the range of the Dreams variable in the Input Y Range window. First click in the **Input Y Range** window. Then click in cell **B1** in the worksheet and drag to cell **B20**. If you prefer, you can manually enter **B1:B20**.
- **Input X Range**. Enter the range of the X1 variable in the Input X Range window. First click in the **Input X Range** window. Then click in cell **C1** in the worksheet and drag to cell **C20**. If you prefer, you can manually enter **C1:C20**.
- **Labels**. Click in the box next to **Labels** to place a check mark there. Cells B1 and C1 contain the labels "Dreams" and "X1," respectively. These labels should not be included in the regression analysis. They will, however, be used to label the output.
- **New Worksheet Ply**. Click the button next to **New Worksheet Ply** to place the output on a sheet different from the one that contains the data.
- **Residuals**. Click in the box next to **Residuals**.

5. Click **OK**. The output that you will obtain is shown on the next page.

	A	B	C	D	E	F	G
1	SUMMARY OUTPUT						
2							
3	Regression Statistics						
4	Multiple R	0.555618656					
5	R Square	0.308712091					
6	Adjusted R Square	0.268048096					
7	Standard Error	4.204028569					
8	Observations	19					
9							
10	ANOVA						
11		df	SS	MS	F	Significance F	
12	Regression	1	134.1760234	134.176	7.59178	0.013516432	
13	Residual	17	300.4555556	17.67386			
14	Total	18	434.6315789				
15							
16		Coefficients	Standard Error	t Stat	P-value	Lower 95%	Upper 95%
17	Intercept	9.777777778	1.401342856	6.977434	2.23E-06	6.82120279	12.73435277
18	X1	5.322222222	1.931617825	2.755318	0.013516	1.246864845	9.3975796

22	RESIDUAL OUTPUT		
23			
24	Observation	Predicted Dreams	Residuals
25	1	15.1	5.9
26	2	15.1	2.9
27	3	15.1	-1.1
28	4	15.1	4.9
29	5	15.1	-4.1
30	6	15.1	3.9
31	7	15.1	-7.1
32	8	15.1	-3.1
33	9	15.1	-2.1
34	10	15.1	-0.1
35	11	9.777777778	2.222222222
36	12	9.777777778	4.222222222
37	13	9.777777778	0.222222222
38	14	9.777777778	-1.777777778
39	15	9.777777778	6.222222222
40	16	9.777777778	-4.777777778
41	17	9.777777778	-6.777777778
42	18	9.777777778	-0.777777778
43	19	9.777777778	1.222222222

Interpreting the Output

Regression Statistics

- **Multiple R**. The square root of the ratio of SS Regression to SS Total, interpreted as the strength of the relationship between treatment and number of dreams.

$$R = \sqrt{\frac{SS_{Regression}}{SS_{Total}}} = \sqrt{\frac{134.1760}{434.6316}} = .5556$$

- **R Square**. The ratio of SS Regression to SS Total, interpreted as the proportion of variability in dreams that is explained by treatment.

$$R^2 = \frac{SS_{Regression}}{SS_{Total}} = \frac{134.1760}{434.6316} = .3087$$

- **Adjusted R Square**. R square adjusted to provide a more accurate estimate of the population R^2 value.

- **Standard Error**. Standard error of estimate, interpreted as the average error in predicting number of dreams by means of the regression equation.

$$SE_{\hat{Y}} = \sqrt{S_Y^2(1-R_{Adj}^2)} = \sqrt{\frac{SS_{Total}}{df_{Total}}(1-R_{Adj}^2)} = \sqrt{\frac{434.6316}{18}(1-.2680)} = 4.2041$$

- **Observations**. The number of observations included in the analysis.

ANOVA

- **Regression**. Variability in number of dreams that is explained by treatment condition.

- **Residual**. Variability in number of dreams that is not explained by treatment condition.

- **Total**. Total variability in number of dreams.

- **F**. The F value is obtained by dividing MS Regression by MS Residual. Note that the square root of F is equal to the t-test value obtained for these same data on page 158.

$$\sqrt{F} = \sqrt{7.5918} = 2.7553$$

- **Significance F**. The probability associated with the obtained value of F. Note that this probability (.0135) is identical to the two-tailed probability of obtained t displayed in the t-test output on page 158.

Coefficients

- **Intercept**. The intercept of the regression equation is equal to the mean of treatment group K, in this case, treatment group 2, the Placebo condition.

$$\bar{Y}_2 = 9.7778$$

- **X1**. The X1 coefficient is equal to the difference between the treatment group means.

$$\bar{Y}_1 - \bar{Y}_2 = 15.1 - 9.7778 = 5.3222$$

The regression equation for predicting number of dreams from treatment is

$$\hat{Y} = 9.7778 + 5.3222X1$$

Residual Output

- **Predicted Dreams**. The predicted number of dreams for each subject is equal to the mean of the subject's treatment condition.

- **Residuals**. The residual for each subject is equal to that subject's observed number of dreams minus that subject's treatment condition mean.

Sample Research Problem with Three Groups

For an example of a three-group dummy coding problem, I will use the same data that were used for the one-way ANOVA in Chapter 9. The regression analysis outcome will be identical to that of the one-way ANOVA.

Steps to Follow to Analyze the Sample Research Problem

1. Open worksheet "Ch11_Trial" on the Web site, or enter the data in an Excel worksheet as shown below. Because treatment has three levels, two dummy variables can be defined. Those variables are named X1 and X2 and are placed in columns C and D, respectively.

	A	B	C	D
1	Group	Rating	X1	X2
2	Clean	6	1	0
3	Clean	5	1	0
4	Clean	7	1	0
5	Clean	6	1	0
6	Clean	3	1	0
7	Clean	7	1	0
8	Criminal	1	0	1
9	Criminal	2	0	1
10	Criminal	1	0	1
11	Criminal	3	0	1
12	Criminal	2	0	1
13	Control	4	0	0
14	Control	3	0	0
15	Control	5	0	0
16	Control	6	0	0
17	Control	4	0	0

The X1 and X2 values can be keyed in directly or you can generate them by using IF statements. For the X1 values, activate cell C2 and enter =IF(A2="Clean",1,0). Press [Enter]. Then copy the contents of cell C2 to cells C3 through C17. For the X2 values, activate cell D2 and enter =IF(A2="Criminal",1,0). Press [Enter]. Then copy the contents of cell D2 to cells D3 through D17.

2. Click the **Data** tab near the top of the screen and select **Data Analysis**.

If Data Analysis does not appear as a choice in the Data ribbon, you will need to load the Microsoft Excel ToolPak add-in. Follow the procedure on page 9.

3. In the Analysis Tools window, select **Regression** and click **OK**.

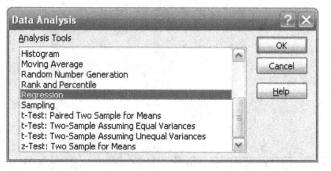

4. Complete the Regression dialog box as shown below. Detailed instructions are given immediately following the dialog box.

- **Input Y Range**. Enter the range of the Rating variable in the **Input Y Range** window. First click in the window. Then click in cell **B1** of the worksheet and drag to cell **B17**. If you prefer, you can manually enter **B1:B17**.
- **Input X Range**. Enter the range of the X1 and X2 variables in the **Input X Range** window. First click in the window. Then click in cell **C1** in the worksheet and drag to cell **D17**. Or, if you prefer, you can manually enter **C1:D17**.
- **Labels**. Click in the box next to **Labels** to place a check mark there. Cells B1, C1, and D1 contain the labels "Rating," "X1," and "X2," respectively, and should not be included in the regression analysis. They will, however, be used to label the output.
- **New Worksheet Ply**. Click in the circle next to **New Worksheet Ply** to place the output in a sheet different from the sheet containing the data.
- **Residuals**. Click in the box next to **Residuals**.

5. Click **OK**. The output that you will obtain is shown on the next page.

	A	B	C	D	E	F	G
1	SUMMARY OUTPUT						
2							
3	*Regression Statistics*						
4	Multiple R	0.826277824					
5	R Square	0.682735043					
6	Adjusted R Square	0.633925049					
7	Standard Error	1.219499687					
8	Observations	16					
9							
10	ANOVA						
11		*df*	*SS*	*MS*	*F*	*Significance F*	
12	Regression	2	41.60416667	20.80208	13.987608	0.000574439	
13	Residual	13	19.33333333	1.487179			
14	Total	15	60.9375				
15							
16		*Coefficients*	*Standard Error*	*t Stat*	*P-value*	*Lower 95%*	*Upper 95%*
17	Intercept	4.4	0.54537684	8.067816	2.039E-06	3.221784969	5.57821503
18	X1	1.266666667	0.738443732	1.715319	0.1100079	-0.328644026	2.86197736
19	X2	-2.6	0.771279324	-3.371023	0.0050139	-4.266247676	-0.9337523

	A	B	C
23	RESIDUAL OUTPUT		
24			
25	*Observation*	*Predicted Rating*	*Residuals*
26	1	5.666666667	0.333333333
27	2	5.666666667	-0.666666667
28	3	5.666666667	1.333333333
29	4	5.666666667	0.333333333
30	5	5.666666667	-2.666666667
31	6	5.666666667	1.333333333
32	7	1.8	-0.8
33	8	1.8	0.2
34	9	1.8	-0.8
35	10	1.8	1.2
36	11	1.8	0.2
37	12	4.4	-0.4
38	13	4.4	-1.4
39	14	4.4	0.6
40	15	4.4	1.6
41	16	4.4	-0.4

Interpreting the Output

Regression Statistics

- **Multiple R**. The square root of the ratio of SS Regression to SS Total, interpreted as the strength of the relationship between treatment and rating.

$$R = \sqrt{\frac{SS_{Regression}}{SS_{Total}}} = \sqrt{\frac{41.6042}{60.9375}} = .8263$$

- **R Square**. The ratio of SS Regression to SS Total, interpreted as the proportion of variability in ratings that is explained by treatment.

$$R^2 = \frac{SS_{Regression}}{SS_{Total}} = \frac{41.6042}{60.9375} = .6827$$

- **Adjusted R Square**. R square adjusted to provide a more accurate estimate of the population R^2 value.

- **Standard Error**. Standard error of estimate, interpreted as the average error in predicting the ratings by means of the regression equation.

$$SE_{\hat{Y}} = \sqrt{S_Y^2(1 - R_{Adj}^2)} = \sqrt{\frac{SS_{Total}}{df_{Total}}(1 - R_{Adj}^2)} = \sqrt{\frac{60.9375}{15}(1 - .6339)} = 1.2195$$

- **Observations**. The number of observations included in the analysis.

ANOVA

- **Regression**. Variability in the ratings that is explained by treatment condition.

- **Residual**. Variability in the ratings that is not explained by treatment condition.

- **Total**. Total variability in the ratings.

- **F**. The F value is obtained by dividing MS Regression by MS Residual. Note that this F of 13.9876 is equal to the one-way ANOVA F-test value obtained for these same data on page 173.

- **Significance F**. The probability associated with the obtained value of F. Note that this probability (.0005) is identical to the probability of obtained F displayed in the ANOVA output on page 173.

Coefficients

- **Intercept**. The intercept of the regression equation is equal to the mean of treatment group K, in this case, group 3, the Control treatment group.

$$\overline{Y}_3 = 4.4$$

- **X1**. The X1 coefficient is equal to the difference between the means of group 1 (Clean) and group 3 (Control).

$$\overline{Y}_1 - \overline{Y}_3 = 5.6667 - 4.4 = 1.2667$$

- **X2**. The X2 coefficient is equal to the difference between the means of group 2 (Criminal) and group 3 (Control).

$$\overline{Y}_2 - \overline{Y}_3 = 1.8 - 4.4 = -2.6$$

The regression equation for predicting rating from treatment is given by

$$\hat{Y} = 4.4 + 1.2667 X1 + (-2.6) X2$$

Residual Output

- **Predicted Rating.** The predicted rating for each subject is equal to the mean of the subject's treatment condition.

- **Residuals.** The residual for each subject is equal to that subject's rating minus that subject's treatment condition mean.

► Section 11.4   Curvilinear Regression

When the relationship between a predictor variable *(X)* and the outcome variable *(Y)* is nonlinear, you may want to carry out a regression analysis to investigate a polynomial relationship between *X* and *Y*. In this section, I explain how to use Excel's procedures to apply the model shown below.

$$\hat{Y} = B_0 + B_1 X + B_2 X^2$$

Sample Research Problem

A statistics instructor was interested in finding out if there was a curvilinear relationship between statistics anxiety and performance on the statistics final exam. The instructor recorded the final exam score and an anxiety measure for each student enrolled the class.

Steps to Follow to Analyze the Sample Research Problem

1. Open worksheet "Ch11_Final Exam" on the Web site, or enter the data in an Excel worksheet.

	A	B	C
1	Final Exam	Anxiety	Anxiety Squared
2	19	5	25
3	18	7	49
4	21	10	100
5	23	12	144
6	28	15	225
7	38	16	256
8	37	17	289
9	36	16	256
10	20	18	324
11	16	19	361
12	14	20	400
13	8	21	441

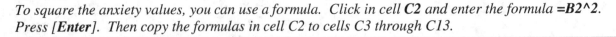

*To square the anxiety values, you can use a formula. Click in cell **C2** and enter the formula =**B2^2**. Press [**Enter**]. Then copy the formulas in cell C2 to cells C3 through C13.*

2. Click the **Data** tab near the top of the screen and select **Data Analysis**.

If Data Analysis does not appear as a choice in the Data ribbon, you will need to load the Microsoft Excel ToolPak add-in. Follow the procedure on page 9.

---

3. In the Data Analysis dialog box, select **Regression** and click **OK**.

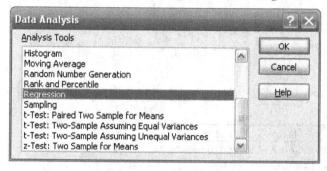

4. Complete the Regression dialog box as shown below.

5. Click **OK**. The summary output provided by Excel's regression procedure is shown below.

	A	B	C	D	E	F	G
1	SUMMARY OUTPUT						
2							
3	*Regression Statistics*						
4	Multiple R	0.72172889					
5	R Square	0.520892591					
6	Adjusted R Square	0.414424278					
7	Standard Error	7.382007286					
8	Observations	12					
9							
10	ANOVA						
11		*df*	*SS*	*MS*	*F*	*Significance F*	
12	Regression	2	533.2203826	266.6102	4.892466	0.036471027	
13	Residual	9	490.4462841	54.49403			
14	Total	11	1023.666667				
15							
16		*Coefficients*	*Standard Error*	*t Stat*	*P-value*	*Lower 95%*	*Upper 95%*
17	Intercept	-19.45950078	15.59009378	-1.2482	0.243454	-54.7267431	15.8077415
18	Anxiety	7.891023461	2.592689872	3.043566	0.013937	2.025951497	13.7560954
19	Anxiety Squared	-0.305681575	0.097925916	-3.12156	0.012288	-0.52720539	-0.08415776

	A	B	C
23	RESIDUAL OUTPUT		
24			
25	*Observation*	*Predicted Final Exam*	*Residuals*
26	1	12.35357715	6.646422848
27	2	20.79926628	-2.799266277
28	3	28.88257634	-7.882576344
29	4	31.21463397	-8.214633974
30	5	30.1274968	-2.127496797
31	6	28.54239144	9.457608562
32	7	26.34592293	10.65407707
33	8	28.54239144	7.457608562
34	9	23.53809127	-3.538091273
35	10	20.11889647	-4.118896466
36	11	16.08833851	-2.088338509
37	12	11.4464174	-3.446417403

Interpreting the Output

Regression Statistics.

- **Multiple R**. The multiple correlation, R, between the two predictor variables (Anxiety and Anxiety-squared) and Y (Final Exam) is .7217.

- **R Square**. R^2 is equal to .5209. Approximately 52 percent of the variation in Final Exam is explainable by the polynomial regression model.

- **Adjusted R Square**. This is an adjusted value that more closely reflects the degree of fit in the population. For this sample problem, adjusted R^2 is equal to .4144.

- **Standard Error**. The standard error of estimate is equal to 7.3820. This is the average amount by which we err if we use our polynomial regression equation to predict Final Exam.

- **Observations**. The data set includes data for 12 statistics students.

ANOVA

- **Regression**. The variation in Final Exam scores that is explained by its relationship with the predictor variables is referred to as *regression*.

- **Residual**. The variation in Final Exam scores that is not explained by the predictor variables is referred to as *residual*.

- **Total**. The total variation in Final Exam scores is partitioned into regression variation and residual variation.

The F value is obtained by dividing MS regression by MS residual. This F test value would be considered statistically significant with alpha equal to .05, because the P-value (.0365) is less than .05.

Coefficient Summary Information

- **Intercept**. The intercept of the regression equation is –19.4595.

- **Anxiety beta weight**. The beta weight for the Anxiety score is 7.8910.

- **Anxiety-squared beta weight**. The beta weight for the curvilinear component, the squared Anxiety score, is –0.3057.

When these values are placed in the regression equation, we see that the polynomial equation for predicting Final Exam from Anxiety is given by

$$\hat{Y} = -19.4595 + 7.8910X - 0.3057X^2$$

If we use the data for the first person in the data set, we obtain

Predicted Final Exam $= -19.4595 + 7.8910(5) - 0.3057(5^2) = 12.35$

- The **Standard Error** values, **t Stat**, **P-value**, and **Lower** and **Upper 95%** confidence interval boundaries are also provided for each of these coefficients. The t Stat for the intercept provides a statistical test of the hypothesis that the intercept is equal to zero. The t tests for the predictor variable beta weights both test the hypothesis that the beta weight is equal to zero. Of special interest here, is the t test for the beta weight of the squared Anxiety measure, because it is this beta weight that reflects the curvilinear component. For this example, the curvilinear component is statistically significant with a P-value equal to .0123.

Residuals

- **Observation**. Refers to the order in which the observations appear in the data set. For example, observation 1 is the first observation in the data set, observation 2 is the second observation in the data set, and so on.

- **Predicted Final Exam**. The value we obtain when we use the polynomial regression equation to predict a student's Final Exam score. The predicted Final Exam score for observation 1 is 12.35.

- **Residuals**. Residuals found by subtracting the predicted value from the actual value. For observation 1, $19 - 12.3536 = 6.6464$.

Scatterplot of a Curvilinear Relation

Excel's chart features enable us to produce a scatterplot with a trendline and also to show the regression equation and R^2 on the graph.

1. Return to the "Ch11_Final Exam" worksheet that has the data for the curvilinear regression example problem. Copy the Anxiety and Final Exam data to a new worksheet so that the columns appear exactly as shown below. Then click in any cell containing data.

Data in column A will be treated as the X variable in the chart and data in column B will be treated as the Y variable.

	A	B
1	Anxiety	Final Exam
2	5	19
3	7	18
4	10	21
5	12	23
6	15	28
7	16	38
8	17	37
9	16	36
10	18	20
11	19	16
12	20	14
13	21	8

2. Click **Insert** near the top of the screen and select **Scatter** in the Charts Group.

3. Select the leftmost diagram in the top row.

4. Let's move the chart to a new worksheet. **Right-click** in the blank space near a border and select **Move Chart**.

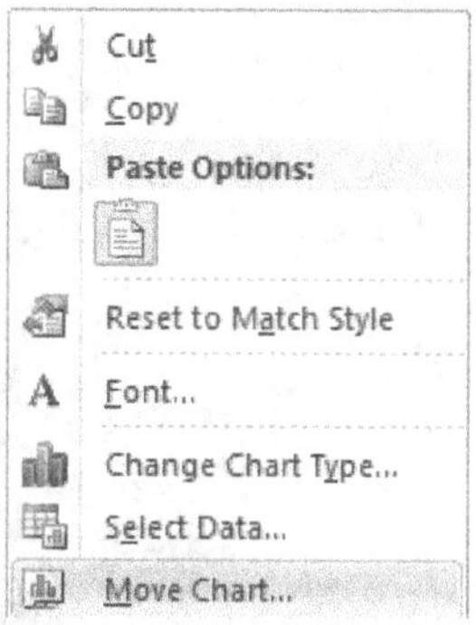

5. Select **New sheet** and click **OK**.

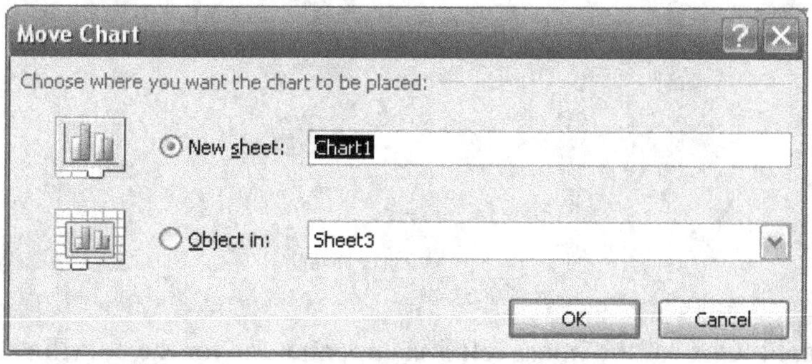

6. Click directly on **Final Exam** at the top of the chart and replace it with **Prediction of Exam Performance from Anxiety**.

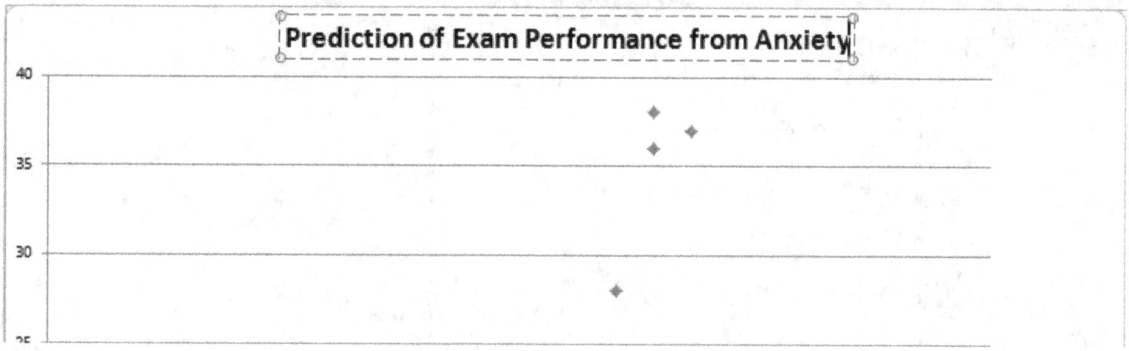

7. Next, let's add axis titles. Click **Axis Titles** in the Layout ribbon.

8. Select **Primary Horizontal Axis Title** and **Title Below Axis**.

9. Replace **Axis Title** with **Anxiety**.

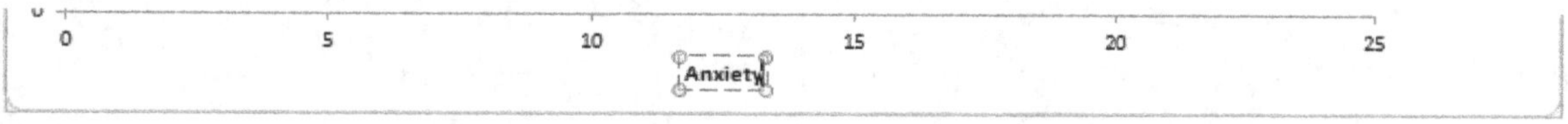

10. Click **Axis Titles** again, and this time select **Primary Vertical Axis Title** and **Rotated Title**.

11. Replace **Axis Title** on the left side of the chart with **Final Exam Score**.

12. Click **Legend** in the Layout ribbon and select **None**.

> None
> Turn off Legend

13. **Right-click** directly on one of the dots in the scatterplot. Select **Add Trendline** from the menu.

> Delete
> Reset to Match Style
> Change Series Chart Type...
> Select Data...
> 3-D Rotation...
> Add Data Labels
> Add Trendline...
> Format Data Series...

14. We will place two trendlines on the graph: One for the linear relationship, and one for the curvilinear relationship. Select **Linear** and click **Close**.

> **Format Trendline**   ? ✕
>
> Trendline Options   Trendline Options
> Line Color   Trend/Regression Type
> Line Style   ○ Exponential
> Shadow   ⦿ Linear
> Glow and Soft Edges   ○ Logarithmic
>   ○ Polynomial   Order: 2
>   ○ Power
>   ○ Moving Average Period: 2

15. **Right-click** directly on one of the dots again and select **Add Trendline** from the menu.

16. Select the **Polynomial** type of Order **2**. At the bottom of the dialog box, select **Display equation on chart** and **Display R-squared value on chart**. Click **Close**.

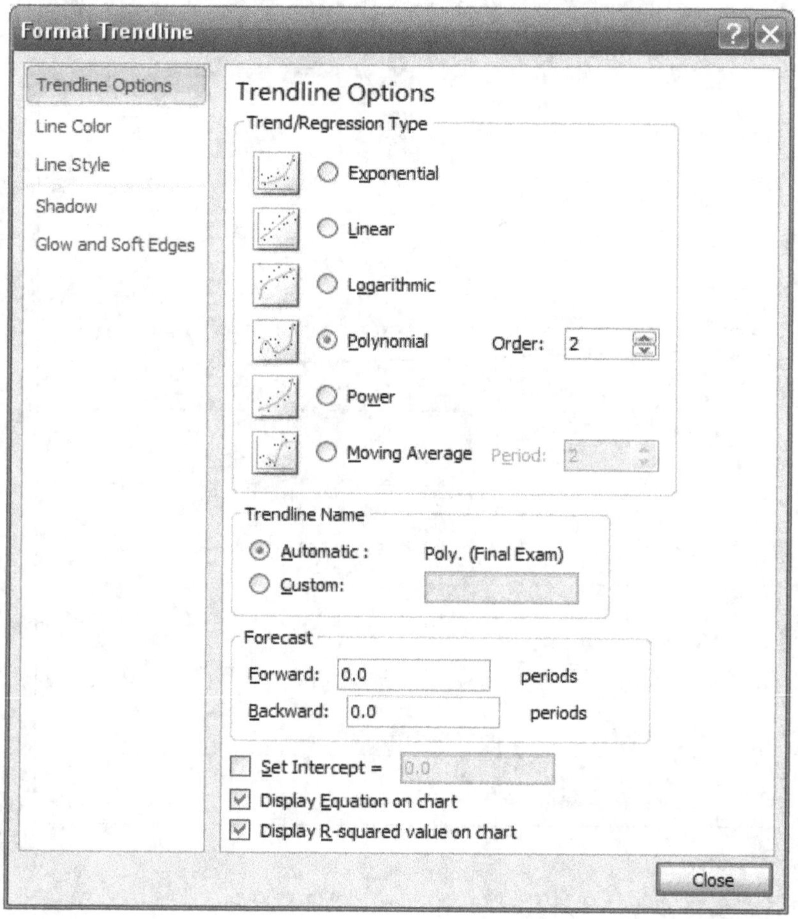

Format Trendline   ? ☒

Trendline Options
Line Color
Line Style
Shadow
Glow and Soft Edges

Trendline Options

Trend/Regression Type

○ Exponential

○ Linear

○ Logarithmic

◉ Polynomial   Order: 2

○ Power

○ Moving Average   Period: 2

Trendline Name

◉ Automatic :   Poly. (Final Exam)

○ Custom:

Forecast

Forward: 0.0   periods

Backward: 0.0   periods

☐ Set Intercept = 0.0

☑ Display Equation on chart

☑ Display R-squared value on chart

Close

The completed graph is shown below.

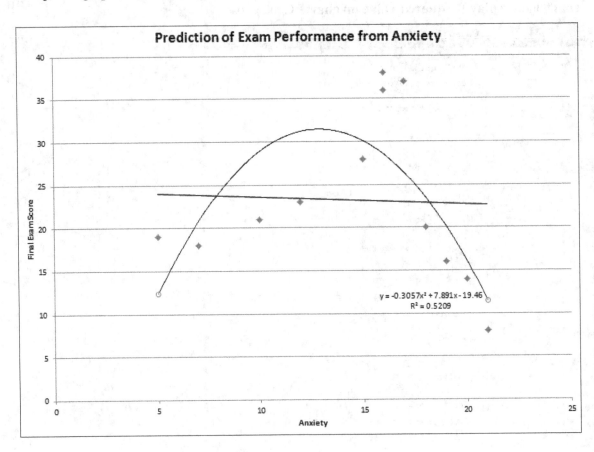

Prediction of Exam Performance from Anxiety

$y = -0.3057x^2 + 7.891x - 19.46$

$R^2 = 0.5209$

| ▶ Section 12.1 | ## Cross Tabulations Using the Pivot Table |

Excel's Pivot Table provides an easy way to cross tabulate both qualitative and quantitative variables. In this chapter, I present three different types of cross tabulation examples: 1) two qualitative variables, 2) one qualitative and one quantitative variable, and 3) three variables (two qualitative and one quantitative). In addition, I show how missing values are displayed.

Sample Research Problem

A survey research project was carried out to gather information about people's attitudes regarding capital punishment. "For" was recorded if respondents indicated they were in favor of capital punishment, and "Against" was recorded if respondents indicated they were against capital punishment. The survey instrument also included some items regarding the participants' demographic characteristics: sex, socioeconomic status (SES), and age. One of the participants did not provide a response for SES, so that cell was left blank. The worksheet containing the data is shown below.

	A	B	C	D
1	OPINION	SEX	SES	AGE
2	Against	Male	Hi	31
3	For	Male	Med	39
4	Against	Female	Med	40
5	Against	Male	Med	51
6	For	Female	Med	40
7	For	Female	Med	42
8	For	Female		40
9	Against	Male	Med	41
10	For	Female	Lo	31
11	Against	Male	Lo	39
12	Against	Female	Lo	54
13	For	Female	Hi	51
14	Against	Male	Med	39
15	Against	Female	Hi	40
16	For	Male	Hi	55
17	For	Male	Lo	32
18	Against	Female	Med	39

Cross Tabulation of Two Qualitative Variables

To illustrate how to cross tabulate two qualitative variables, we will work with OPINION regarding capital punishment and SEX of the respondent.

1. Open worksheet "Ch12_Opinion" on the Web site or enter the data in an Excel worksheet as shown on the preceding page. Click in any cell containing data before you start the Pivot Table procedure so that the data range will be automatically entered in the Table/Range window.

2. Click the **Insert** tab near the top of the screen and select **Pivot Table**.

3. You are given the options of Pivot Table or Pivot Chart. Select **Pivot Table**.

4. You should see Sheet1!$A$1:$D$18 in the Table/Range window of the Create Pivot Table dialog box. Sheet1 is the name of the worksheet and $A$1:$D$18 is the range that includes the data for all four variables. For this example, we will place the summary table in a new worksheet. To do this, just select **New Worksheet** in the lower part of the dialog box. Click **OK**.

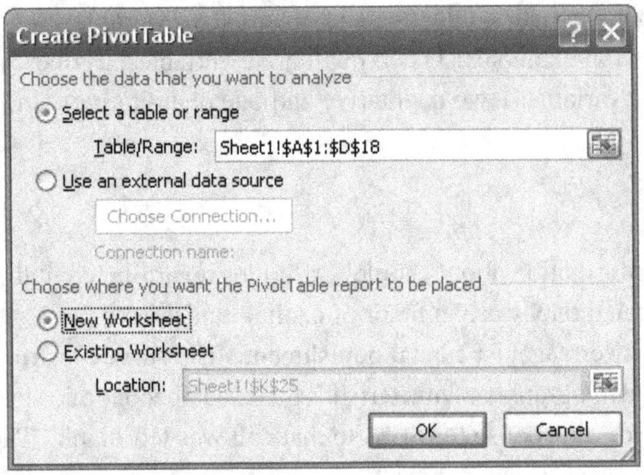

5. On the right side of the worksheet, you will see a Pivot Table Field List displaying OPINION, SEX, SES, and AGE. These are the labels that appeared in the top cells (A1, B1, C1, and D1) of the data range. The field list shown below also displays Report Filter, Column Labels, Row Labels, and Σ Values. To make it easy for you to follow along, you will want to use the same display. Click the down arrow in the space below Pivot Table Field List and select **Field Section and Areas Section Side-by-Side**.

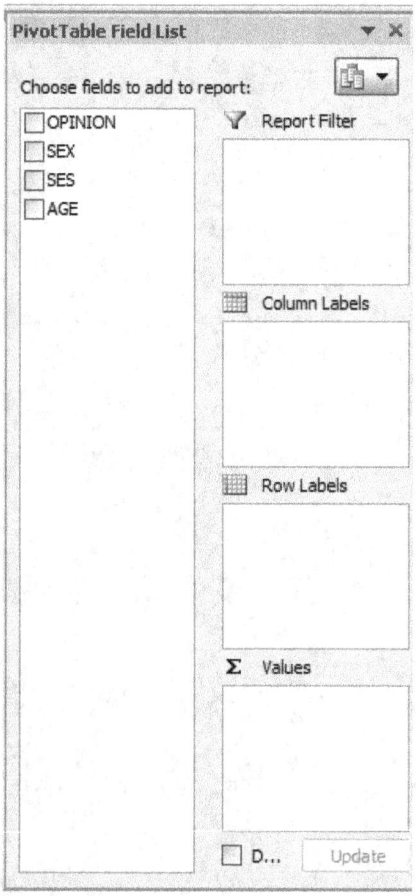

6. We will be cross tabulating OPINION by SEX. To select these variables, click in the boxes next to the variable labels.

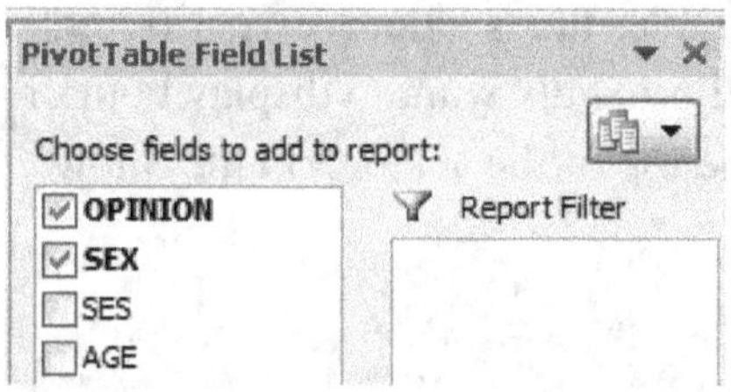

7. Depending on your preferences, you can place OPINION on either the rows or the columns of the cross tabulation. For this example, we will place OPINION on the columns and SEX on the rows. Then, to obtain a count of OPINION responses cross tabulated by SEX, we will place OPINION on the data area of the table. Click directly on **OPINION** in field list on the left and drag it to the **Column Labels** section. Click **SEX** in the field list on the left and drag it to the **Row Labels** section. Click **OPINION** in the field list again and drag it to the Σ **Values** section.

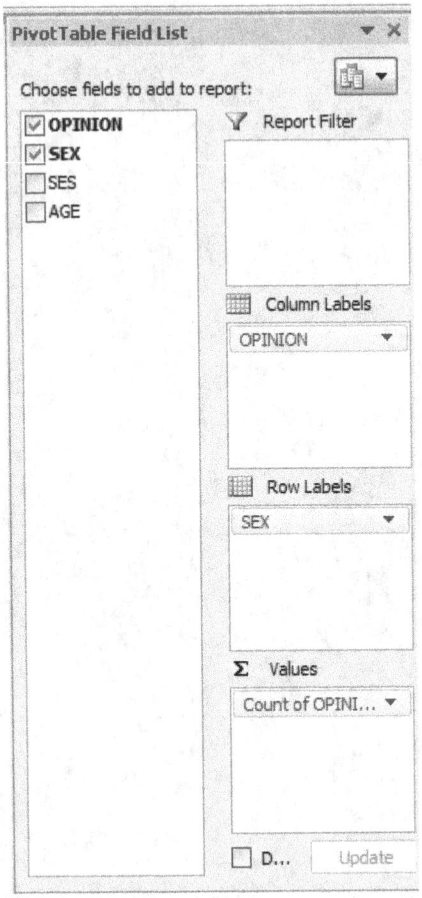

8. Your cross tabulation summary table should look like the one shown below. The down arrows next to OPINION and SEX allow you to select specific values of these variables.   For example, you may want to display only the females' responses. Because each of these variables takes on only two different values (e.g., "Female" and "Male"; "Against" and "For"), you will probably want to display both of them. Let's get rid of these arrows. Click the **Options** tab near the top of the screen. In the Show group, click **Field List** and **Field Headers**.

	A	B	C	D
1	Drop Report Filter Fields Here			
2				
3	Count of OPINION	OPINION ▼		
4	SEX ▼	Against	For	Grand Total
5	Female	4	5	9
6	Male	5	3	8
7	Grand Total	9	8	17

9. Click in a cell of the table so that the table is active and can be revised. Let's select a more professional appearance for the table. Click the **Design** tab near the top of the screen. Then click the down arrow on the right side of the Pivot Table Styles group.   A variety of color and style options are available. Let's select the leftmost style in the top row under Medium.

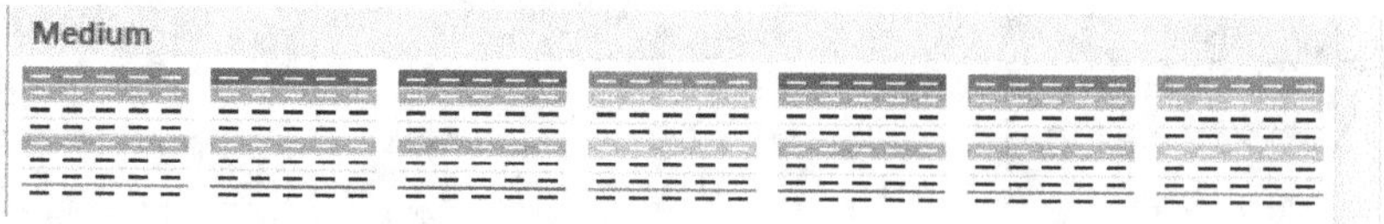

Medium

10. Excel automatically displays values in alphabetical order. Let's switch the order in which the OPINION responses are displayed. **Right-click** on **Against** in cell B4 and select **Move** and **Move "Against" Right**.

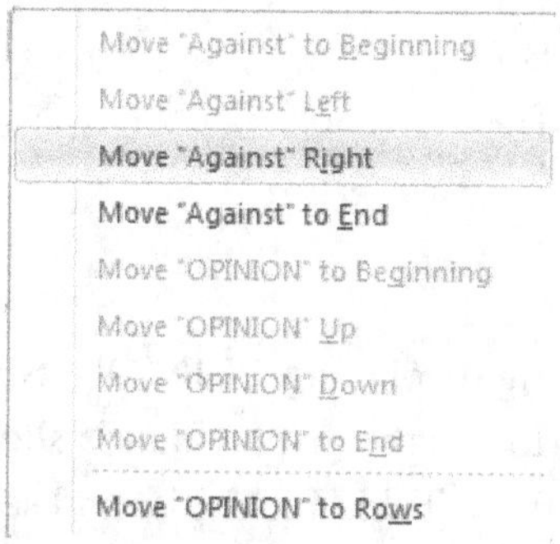

Move "Against" to Beginning
Move "Against" Left
Move "Against" Right
Move "Against" to End
Move "OPINION" to Beginning
Move "OPINION" Up
Move "OPINION" Down
Move "OPINION" to End
Move "OPINION" to Rows

The completed table is shown below.

	A	B	C	D
1				
2				
3	Count of OPINION			
4		For	Against	Grand Total
5	Female	5	4	9
6	Male	3	5	8
7	Grand Total	8	9	17

Cross Tabulation of a Qualitative and a Quantitative Variable

For my illustration of cross tabulating a qualitative variable by a quantitative variable, I will utilize the responses given to **OPINION** and **AGE**.

Before beginning this example, the data must be entered in an Excel worksheet. Open worksheet "Ch12_Opinion" on the Web site. Or, if you haven't already done so, enter the data shown on the first page of this chapter into an Excel worksheet.

1. The worksheet containing the data (most likely Sheet 1) should be displayed on your screen. If it is not, return to that worksheet now. To do this, click the Sheet 1 tab near the bottom of the screen.

2. Click in any cell containing data. Then click **Insert** near the top of the screen and select **Pivot Table**. You can select a Pivot Table or a Pivot Chart. Select **Pivot Table**.

3. You should see Sheet1!$A$1:$D$18 in the Table/Range window. The placement option **New worksheet** has been preselected. Click **OK**.

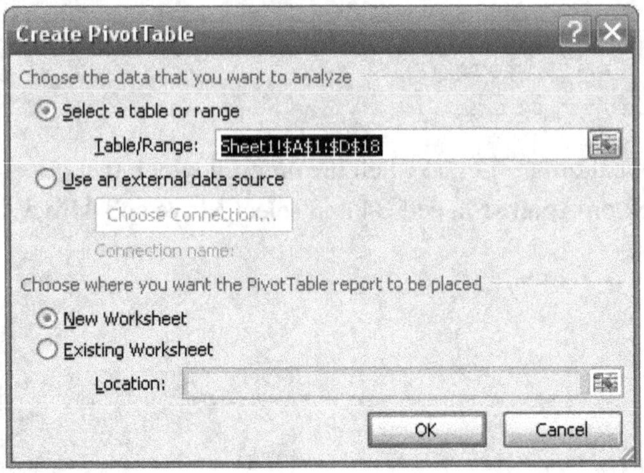

4. On the right side of the worksheet, you will see a Pivot Table Field Listing displaying OPINION, SEX, SES, and AGE. To make it easy for you to follow along, you will want to use the display that is shown below. To select this display, click the down arrow in the space below Pivot Table Field List and select **Field Section and Areas Section Side-by-Side**.

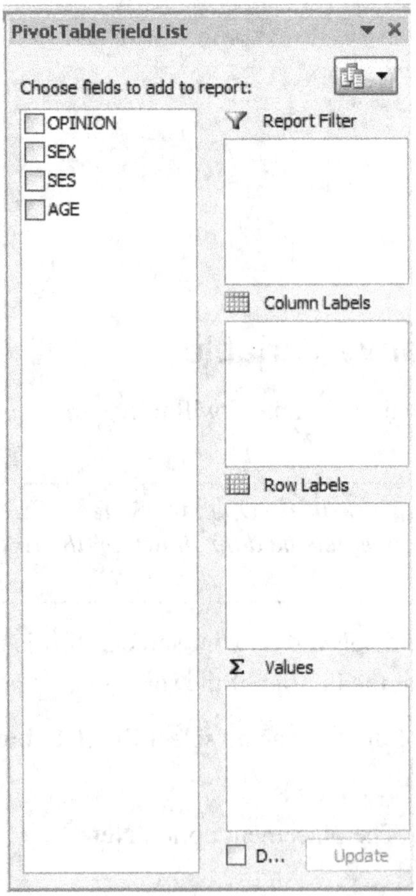

5. We will be cross tabulating OPINION by AGE. To select these variables, click in the boxes to the left of the variable labels.

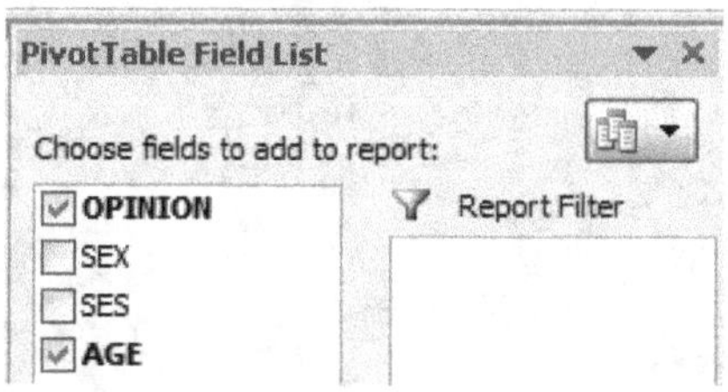

6. OPINION is recorded as either "For" or "Against." AGE, however, takes on several values from 31 to 55. Therefore, the table will be more attractive and will more easily fit on one page if you place AGE on the rows and OPINION on the columns. Click directly on **OPINION** in the field list on the left and drag it to the **Column Labels** section. Click **AGE** in the field list on the left and drag it to the **Row Labels** section. If AGE already appears in the Σ Values section, you don't need to do anything further. Otherwise, click **AGE** in the field list again and drag it to the **Σ Values** section. Your display should look like the one shown below.

When you are working with two qualitative variables, it does not matter a great deal which one you drag to the data (Σ Values) area of the Pivot Table. Quantitative variables, however, may require an extra step that you should be aware of, so this example will give you practice working with a quantitative variable (AGE).

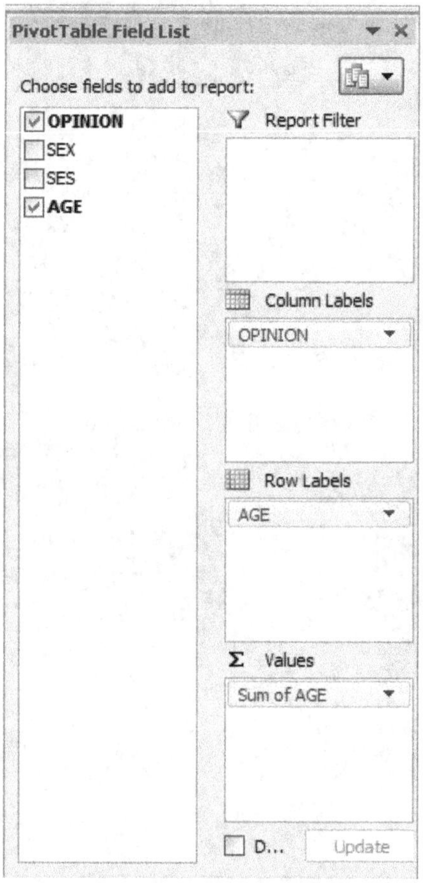

7. You see **Sum of AGE** in the Σ Values section because the default summary measure for a quantitative variable is sum. You want count of age, not sum of age. To change this, click the down arrow to the right of **Sum of Age** and select **Value Field Settings**.

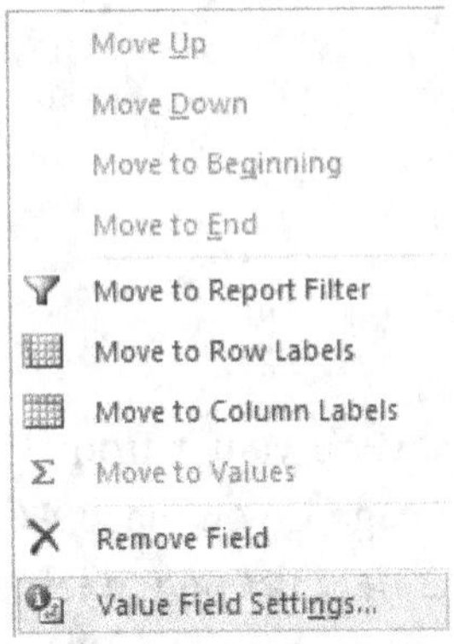

```
        Move Up
        Move Down
        Move to Beginning
        Move to End
   ▽    Move to Report Filter
   ▦    Move to Row Labels
   ▦    Move to Column Labels
   Σ    Move to Values
   ✕    Remove Field
   ⓘ    Value Field Settings...
```

8. In the **Summarize value field by** window, select **Count** and click **OK**. .

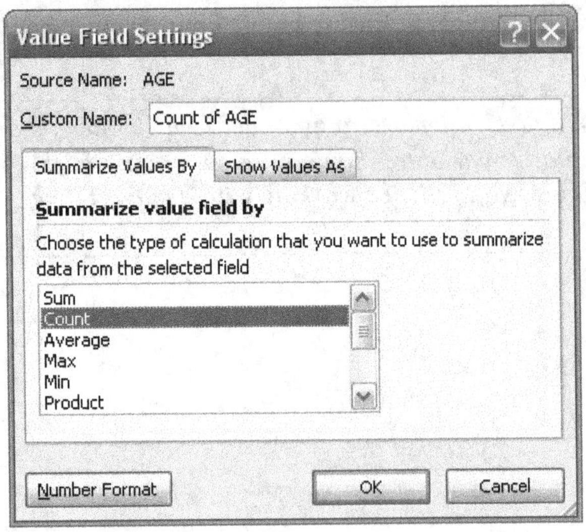

Value Field Settings   ? ✕

Source Name: AGE

Custom Name: Count of AGE

| Summarize Values By | Show Values As |

Summarize value field by

Choose the type of calculation that you want to use to summarize
data from the selected field

```
Sum
Count
Average
Max
Min
Product
```

Number Format   OK   Cancel

9. The table shown below displays counts rather than sums. The table is not quite finished, however, because there are a number of empty cells. You would like to see zeroes in the empty cells. **Right-click** in a cell within the table.

	A	B	C	D
1	Drop Report Filter Fields Here			
2				
3	Count of AGE	OPINION ▾		
4	AGE ▾	Against	For	Grand Total
5	31	1	1	2
6	32		1	1
7	39	3	1	4
8	40	2	2	4
9	41	1		1
10	42		1	1
11	51	1	1	2
12	54	1		1
13	55		1	1
14	Grand Total	9	8	17

10. Select **Pivot Table Options** from the menu that appears.

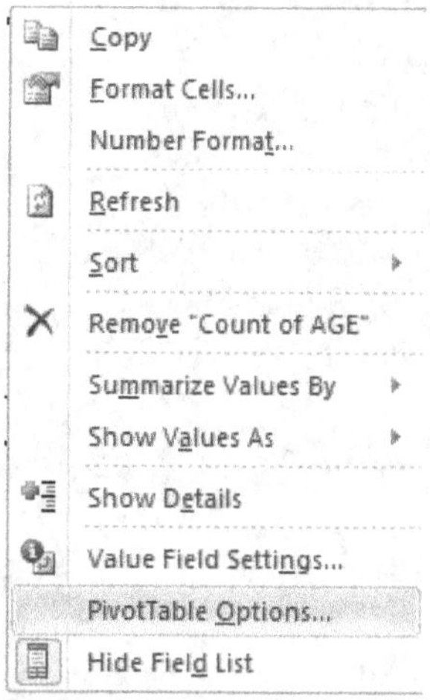

- Copy
- Format Cells...
- Number Format...
- Refresh
- Sort ▸
- ✕ Remove "Count of AGE"
- Summarize Values By ▸
- Show Values As ▸
- Show Details
- Value Field Settings...
- PivotTable Options...
- Hide Field List

11. Click in the box next to **For empty cells show** to place a check mark there. Then enter a 0 in the box. Click **OK**.

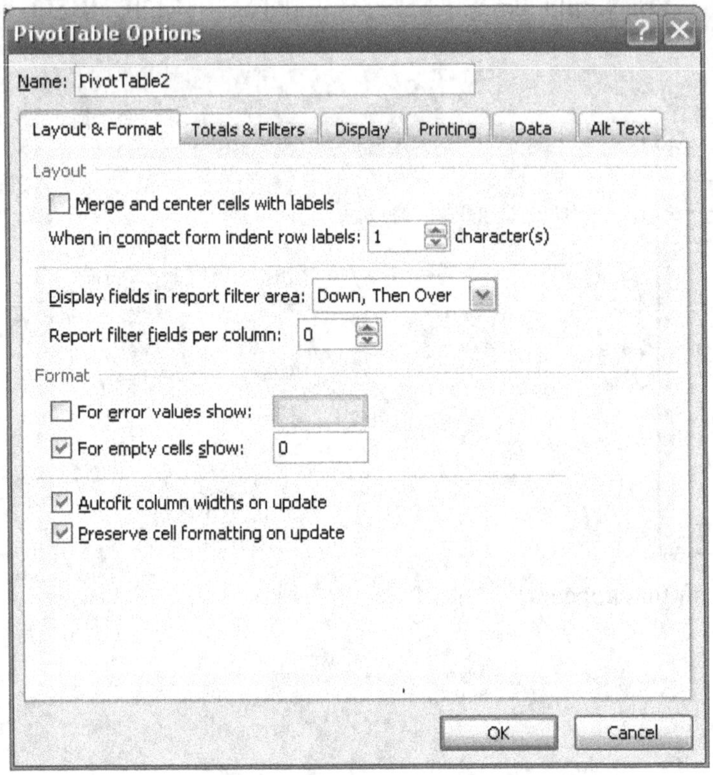

The final version of the table is shown below.

	A	B	C	D
1				
2				
3	Count of AGE	OPINION		
4	AGE	Against	For	Grand Total
5	31	1	1	2
6	32	0	1	1
7	39	3	1	4
8	40	2	2	4
9	41	1	0	1
10	42	0	1	1
11	51	1	1	2
12	54	1	0	1
13	55	0	1	1
14	Grand Total	9	8	17

Cross Tabulation of Three Variables

For a cross tabulation example with three variables, we will use OPINION, SEX, and AGE. Because OPINION and SEX both take on only two different values and AGE takes on several values, we will place AGE on the rows of the table and both OPINION and SEX on the columns. This layout generates a table that is easier to read and use than if we request the opposite placement of the variables on the rows and columns of the table. I suggest that you experiment with different layouts to find the presentation that best fits your needs.

It really does not matter if you place OPINION or SEX in the data area of the table. What does matter is the order in which OPINION and SEX are placed in the column dimension of the table. I will show the output that is obtained with the two different orders.

Before beginning this example, the data must be entered in an Excel worksheet. If you have not already done so, open worksheet "Ch12_Opinion" on the Web site. Or enter the data shown on the first page of this chapter into an Excel worksheet.

1. First return to the worksheet that contains the data (most likely Sheet 1). To do this, click the Sheet 1 tab near the bottom of the worksheet.

2. Click in any cell of the table that contains data. Then click the **Insert** tab near the top of the screen and select **Pivot Table**. Select the **Pivot Table** option.

3. You should see Sheet1!$A$1:$D$18 in the Table/Range window. (If you need to enter the range, you can type **A1:D18** in the window.) For placement of the Pivot Table, the option **New Worksheet** should already be selected. Click **OK**.

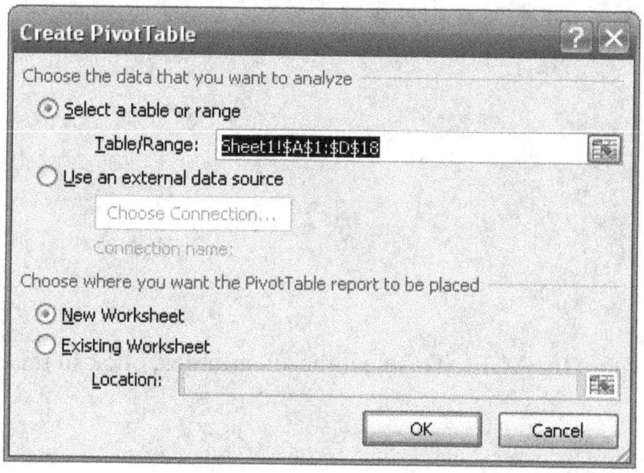

4. If your display is not the same as the one shown below, you will want to change it so that it is easier for you to follow along.   To select this display, click the down arrow in the space below Pivot Table Field List and select **Field Section and Areas Section Side-by-Side**.

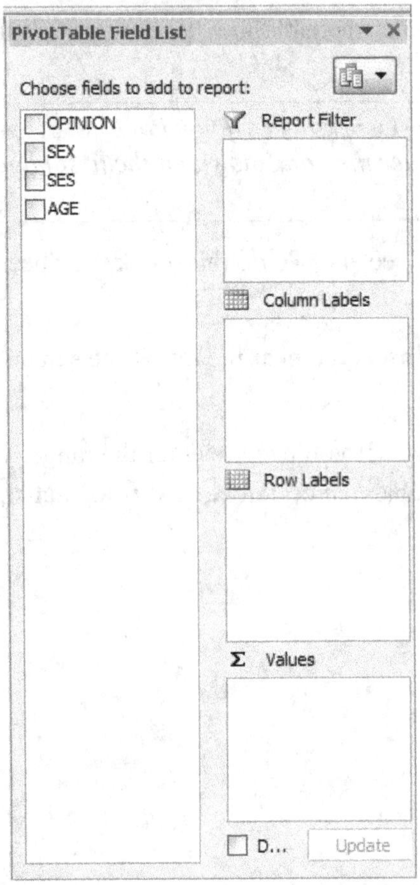

5. We will be cross tabulating OPINION by SEX and by AGE. To select these variables, click in the boxes to the left of the variables labels.

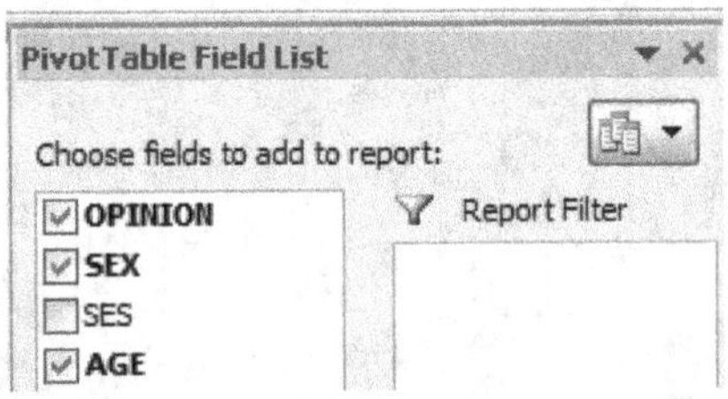

6. For this version of the table, we will place OPINION on the columns first, followed by SEX. In both versions, AGE will be placed on the rows, and OPINION will be placed in the data area. Click directly on **OPINION** in the field list on the left and drag it to the **Column Labels** section. Click **SEX** in the field list and drag it to the **Column Labels** section. Click **AGE** in the field list and drag it to the **Row Labels** section. Click **OPINION** in the field list and drag it to the Σ **Values** section. If you make a mistake or change your mind, just drag the variables back to the field list. Your field list and areas should look the same as in the display shown at the top of the next page.

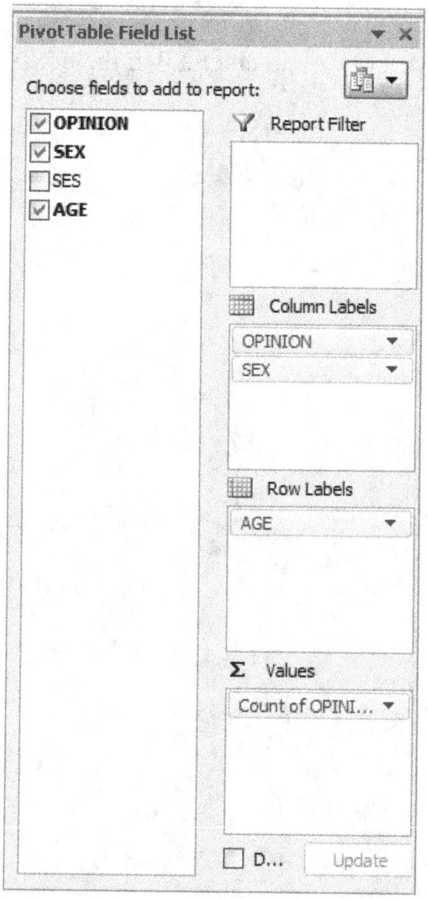

7. If you would like to see zeroes in the blank cells, **right-click** in any cell of the table and select **Pivot Table Options** from the menu that appears.

	A	B	C	D	E	F	G	H
1		Drop Report Filter Fields Here						
2								
3	Sum of AGE	OPINION ▼	SEX ▼					
4		⊟Against		Against Total	⊟For		For Total	Grand Total
5	AGE ▼	Female	Male		Female	Male		
6	31		31	31	31		31	62
7	32					32	32	32
8	39	39	78	117		39	39	156
9	40	80		80	80		80	160
10	41		41	41				41
11	42				42		42	42
12	51		51	51	51		51	102
13	54	54		54				54
14	55					55	55	55
15	Grand Total	173	201	374	204	126	330	704

8. Click in the box next to **For empty cells show** to place a check mark there and enter a 0 in the box. Click **OK**.

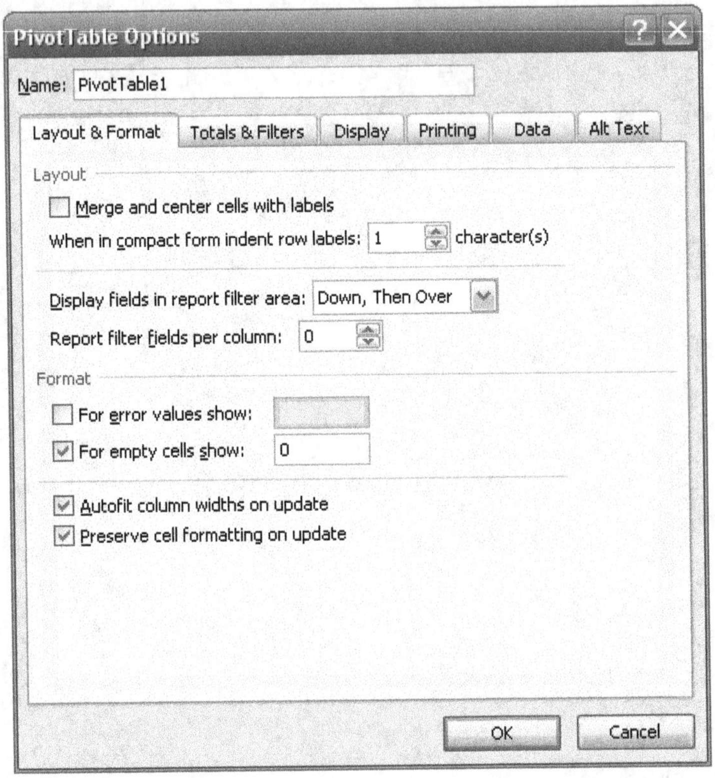

9. The completed table with zeroes displayed in empty cells is shown below. Let's see what the table looks like when we change the order in which SEX and OPINION are placed in the layout. First be sure that the table is active and can be revised. Just click in any cell of the table and you will see the Pivot Table Field list on the right.

	A	B	C	D	E	F	G	H
1								
2								
3	Sum of AGE	OPINION ▼	SEX ▼					
4		⊟ Against		Against Total	⊟ For		For Total	Grand Total
5	AGE ▼	Female	Male		Female	Male		
6	31	0	31	31	31	0	31	62
7	32	0	0	0	0	32	32	32
8	39	39	78	117	0	39	39	156
9	40	80	0	80	80	0	80	160
10	41	0	41	41	0	0	0	41
11	42	0	0	0	42	0	42	42
12	51	0	51	51	51	0	51	102
13	54	54	0	54	0	0	0	54
14	55	0	0	0	0	55	55	55
15	Grand Total	173	201	374	204	126	330	704

10. To make the modification, just click on the **SEX** field button in the Column Labels area at the right and drag it on top of the OPINION field button that is above it.

The contents of the table will be re-tabulated with SEX placed before OPINION on the columns.

	A	B	C	D	E	F	G	H
1								
2								
3	Sum of AGE	SEX	OPINION					
4		⊟ Female		Female Total	⊟ Male		Male Total	Grand Total
5	AGE	Against	For		Against	For		
6	31	0	31	31	31	0	31	62
7	32	0	0	0	0	32	32	32
8	39	39	0	39	78	39	117	156
9	40	80	80	160	0	0	0	160
10	41	0	0	0	41	0	41	41
11	42	0	42	42	0	0	0	42
12	51	0	51	51	51	0	51	102
13	54	54	0	54	0	0	0	54
14	55	0	0	0	0	55	55	55
15	Grand Total	173	204	377	201	126	327	704

Missing Data

When you use Excel for statistical analysis, missing values should be indicated by blanks in your worksheet. I have only one missing value in my data set which occurs in the SES column. The subject whose data are displayed in row 8 did not provide an SES response. To illustrate how the Pivot Table handles missing values, let's cross tabulate OPINION and SES.

Before beginning this example, the data must be entered in an Excel worksheet. Open worksheet "Ch12_Opinion" on the Web site. Or, if you haven't already done so, enter the data shown on the first page of this chapter into an Excel worksheet.

1. Return to the sheet that contains the data set (most likely Sheet 1). To do this, click the Sheet 1 tab near the bottom of the screen.

2. Click the **Insert** tab near the top of the screen and select **Pivot Table**. Select the **Pivot Table** option.

3. The entries in your Create Pivot Table dialog box should be the same as those shown below. Make any necessary revisions and then click **OK**.

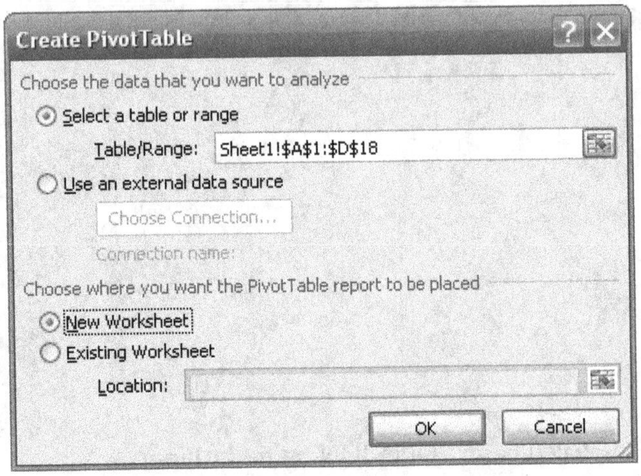

4. Click directly on **SES** in the field list and drag it to the **Column Labels** section. Click **OPINION** in the field list and drag it to the **Row Labels** section. Click **OPINION** in the field list again and drag it to the Σ **Values** section. If you make a mistake or change your mind, just drag the variables back to the field list. Your display should be the same as the one shown below.

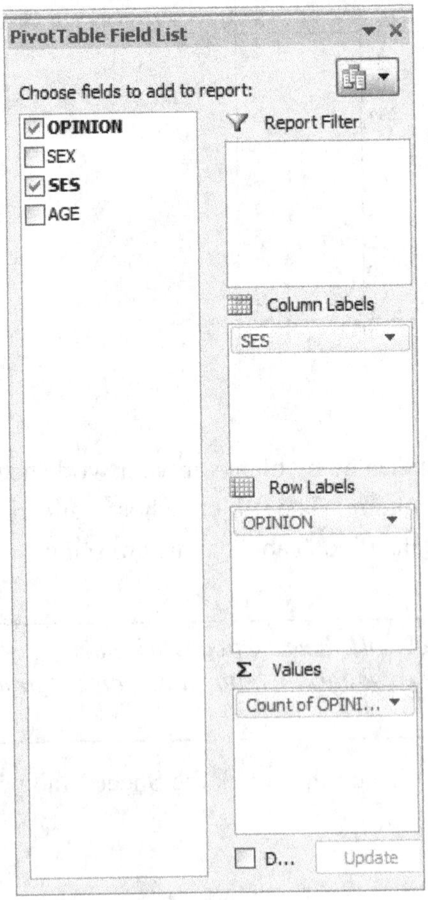

5. The completed cross tabulation is shown below. Notice the column headed "(blank)". This column contains a tally of the OPINION responses for the survey participants who did not answer the SES question.

	A	B	C	D	E	F
1						
2						
3	Count of OPINION	SES ▼				
4	OPINION ▼	Hi	Lo	Med	(blank)	Grand Total
5	Against	2	2	5		9
6	For	2	2	3	1	8
7	Grand Total	4	4	8	1	17

► Section 12.2   Chi-Square Test of Independence

When a researcher is interested in investigating the relationship between qualitative variables, the appropriate test is often the chi-square test of independence. To carry out this test, we first need to calculate the observed and expected frequencies for each cell of the cross tabulation. Although the Pivot Table can be used to provide observed frequencies, we will need to key in our own formulas to calculate the expected frequencies. Once we have the observed and expected frequencies, Excel's functions can be utilized to complete the computations for chi-square and to evaluate its statistical significance.

The formula for the chi-square test of independence is given by

$$X^2 = \sum \frac{\left(O_j - E_j\right)^2}{E_j}$$, where O_j is an observed cell frequency and E_j is an expected cell frequency.

Assumptions Underlying the Chi-Square Test of Independence

Four assumptions underlie the chi-square test of independence:

1. Subjects are randomly and independently sampled from the population of interest.
2. Measures are obtained from a single sample.
3. Variables included in the analysis are measured on a qualitative scale.
4. Expected cell frequencies are greater than or equal to five.

Sample Research Problem

A researcher had a hunch that men and women had different preferences in spectator sports. More specifically, the researcher thought that men preferred watching sports with aggressive physical contact whereas women preferred watching sports without aggressive physical contact. The researcher administered a questionnaire to a random sample of 22 college students. The questionnaire consisted of two items. One item asked which sport the participant enjoyed watching more: football or tennis. The other item asked the participant's gender: female or male.

Using the Pivot Table for Observed Cell Frequencies

1. Open worksheet "Ch12_Preferred Sport" on the Web site, or enter the data in an Excel worksheet as displayed below.

	A	B	C
1	Gender	Preferred Sport	
2	Male	Football	
3	Male	Tennis	
4	Female	Tennis	
5	Male	Tennis	
6	Female	Tennis	
7	Male	Football	
8	Female	Tennis	
9	Female	Football	
10	Male	Football	
11	Female	Tennis	
12	Male	Football	
13	Male	Tennis	
14	Female	Tennis	
15	Male	Football	
16	Male	Football	
17	Female	Tennis	
18	Female	Tennis	
19	Male	Football	
20	Male	Tennis	
21	Female	Tennis	
22	Female	Tennis	
23	Female	Football	

2. Click in any cell that contains data so that the data range will be automatically entered in the Table/Range window of the Pivot Table dialog box. Then click the **Insert** tab near the top of the screen and select **Pivot Table**. Select the **Pivot Table** option.

3. You should see Sheet1!$A$1:$B$23 in the Table/Range window. If you don't, just click in the window and enter **A1:B23**. Select **New Worksheet** for placement of the Pivot Table report. Click **OK**.

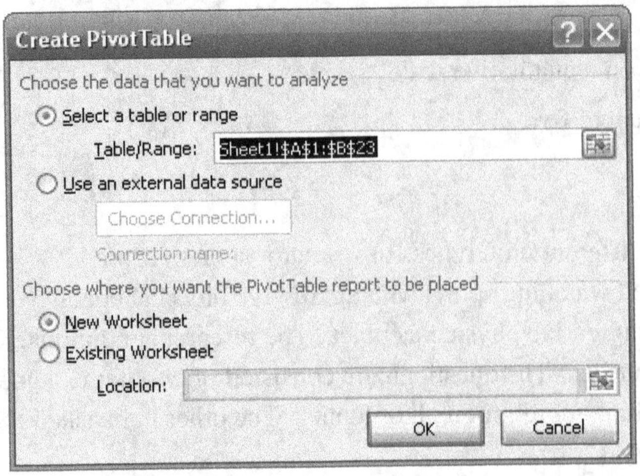

4. On the right side of the worksheet, you will see a Pivot Table Field List displaying Gender and Preferred Sport. These are the labels that appeared in the top cells (A1 and B1) of the data range. To make it easy for you to follow along, you will want to use the same display that is shown below. If your display is not the same, just click the down arrow in the space below Pivot Table Field List and select **Field Section and Areas Section Side-by-Side**.

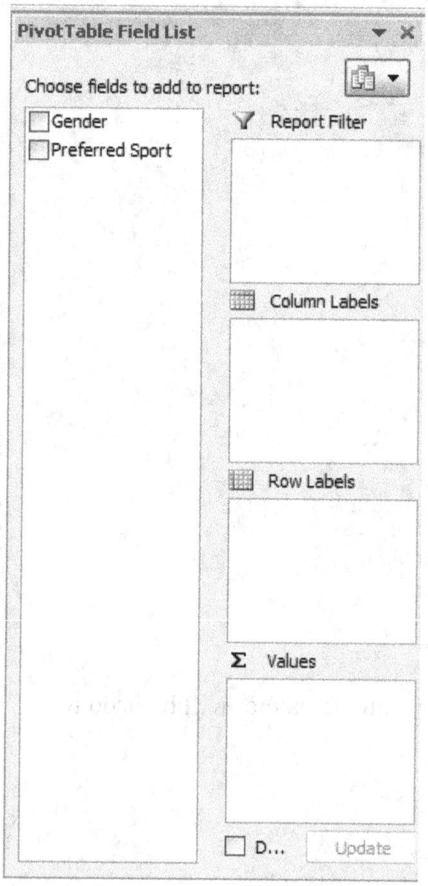

5. We will place Gender on the columns and Preferred Sport on the rows. When you are cross tabulating just two variables, it doesn't matter which of the two variables is placed in the Σ Values section. For this example, let's place Gender in the Σ Values section. Click directly on **Gender** in the field list on the left and drag it to the **Column Labels** section. Click **Preferred Sport** in the field list and drag it to the **Row Labels** section. Click **Gender** in the field list again and drag it to the Σ **Values** section. If you make a mistake, just drag the variable back to the field list. Your worksheet should look similar to the one shown at the top of the next page.

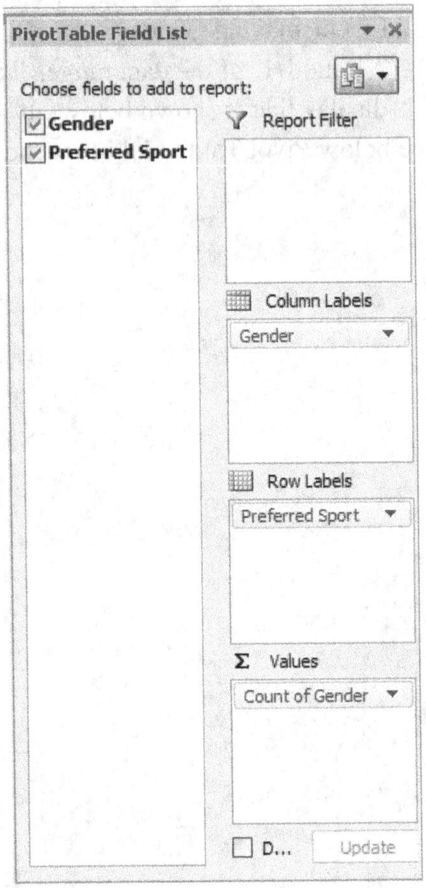

The completed table is shown below. This table contains the observed frequencies that will be used to calculate chi-square.

	A	B	C	D
1				
2				
3	Count of Gender	Column Labels ▼		
4	Row Labels ▼	Female	Male	Grand Total
5	Football	2	7	9
6	Tennis	9	4	13
7	Grand Total	11	11	22

Using Formulas to Calculate Expected Cell Frequencies

You need the observed row and column frequencies to calculate the expected cell frequencies. For convenient access to these observed frequencies, I have placed the cross tabulation table in "Ch12_Preferred Sport" in a sheet labeled "Chi Square." Cells in this sheet contain formulas for calculating expected frequencies and for calculating the chi-square test statistic.   This sheet can serve as a template for carrying out a chi-square test for other cross tabulations where you are working with just two variables that each take on two values.   We'll start out with instructions for calculating the expected cell frequencies.

1. If you have not already done so, open worksheet "Ch12_Preferred Sport" on the Web site. Click the **Chi Square** sheet tab located near the bottom of the screen, or enter the information that you see below

using exactly the same cell locations. The table on the top displays the observed frequencies. These are the same frequencies displayed in the Pivot Table on the preceding page. Cells in the range B12:C13 contain formulas. The formulas are displayed here so that you can see exactly what was entered in each cell. If you would like to display the formulas in your worksheet, click the **Formulas** tab near the top of the screen, and select **Show Formulas** in the Formula Auditing group. To hide the formulas, just click **Show Formulas** again, and you will see the numerical results of the computations that were carried out in the formulas.

▲	A	B	C	D
1				
2				
3				
4		Female	Male	Grand Total
5	Football	2	7	9
6	Tennis	9	4	13
7	Grand Total	11	11	22
8				
9				
10	Expected Frequencies			
11		Female	Male	
12	Football	=B7*D5/D7	=C7*D5/D7	
13	Tennis	=B7*D6/D7	=C7*D6/D7	

The calculation formula for the expected cell frequencies is

$$E_j = \frac{(Column\ Grand\ Total)(Row\ Grand\ Total)}{Overall\ Grand\ Total}$$

where the totals refer to grand totals displayed in the observed frequencies table. Each expected cell frequency is found by multiplying its column grand total by its row grand total and then dividing by the overall grand total (22).

For example, the expected frequency for the Female/Tennis cell would be computed as follows:

$$E = \frac{(11)(13)}{22} = 6.5$$

For efficiency and accuracy, I recommend clicking in cells instead of keying in the information. So, rather than using numerical values, we will be using cell addresses. The formula you will enter in the Female/Tennis cell is **=B7*D6/D7**. A similar formula is used to calculate every expected cell frequency.

2. Enter the four formulas in the range B12:C13 as shown in the worksheet. Press [**Enter**] after entering each formula.

3. When the numerical contents of the cells are displayed rather than the formulas, you should see the numbers that are shown at the top of next page. If you do not see the same result in any of the cells, you will need to correct the formula.

To display formulas, select Show Formulas in the Formula Auditing Group of the Formulas ribbon. To hide formulas, select Show Formulas again.

10	Expected Frequencies		
11		Female	Male
12	Football	4.5	4.5
13	Tennis	6.5	6.5

In the following section, I explain how to use Excel's functions to find the critical value of chi-square for a specified alpha, to find the P-value associated with the observed value of chi-square, and to calculate observed chi-square. The results of the calculations will be placed in the lower part of the worksheet.

Using Functions for a Chi-Square Test of Independence

Now that you have observed and expected frequencies displayed in a worksheet, you can utilize Excel's functions to finish the chi-square calculations for you.

If you are working in the Chi Square sheet in "Ch12_Preferred Sport," all the information should already be displayed. If you need to enter labels and formulas, key in the labels, numerical values, and formulas as shown below in rows 15 to 19. Press [Enter] after each cell entry is completed.

10	Expected Frequencies		
11		Female	Male
12	Football	=B7*D5/D7	=C7*D5/D7
13	Tennis	=B7*D6/D7	=C7*D6/D7
14			
15	Alpha	0.05	
16	df	1	
17	Critical chi-square	=CHISQ.INV.RT(B15,B16)	
18	P-Value	=CHISQ.TEST(B5:C6,B12:C13)	
19	Observed chi-square	=CHISQ.INV.RT(B18,B16)	

Your results should be the same as those shown below. If any of your numbers are different, you will need to correct the cell entries.

15	Alpha	0.05
16	df	1
17	Critical chi-square	3.841458821
18	P-Value	0.030147622
19	Observed chi-square	4.700854701

Interpreting the Output

- **Alpha**. Alpha is the type 1 error probability for the statistical test. The value we selected, .05, is commonly used. Another commonly used value is .01. If you change the alpha value in the worksheet, you will see that critical chi-square also changes.
- **df**. The formula for degrees of freedom (df) for the chi-square test of independence is
  $df = (r - 1)(c - 1)$ where r refers to the number of rows in the cross tabulation and c refers to the number of columns. Because this cross tabulation has 2 rows and 2 columns, $df = (2 - 1)(2 - 1) = 1$.
- **Critical chi-square**. Excel's CHISQ.INV.RT function was used to find the critical chi-square value. CHISQ.INV.RT asks you to provide the type 1 error probability (alpha) and the degrees of freedom. The test result is statistically significant if observed chi-square is greater than critical chi-square.

- **P-value**. The P-value is the probability associated with the observed (or calculated) value of chi-square. Excel's CHISQ.TEST function was used to find the P-value. CHISQ.TEST asks you to provide the worksheet range of the observed cell frequencies and the worksheet range of the expected cell frequencies.

- **Observed chi-square**. Excel's CHISQ.INV.RT function was used to find the observed chi-square value. You used the CHISQ.INV.RT function earlier in this example. Recall that CHISQ.INV.RT returns the chi-square associated with a specified P-value and df. This time you use the P-value rather than alpha for the probability.

- The observed chi-square of 4.70 exceeds the critical chi-square of 3.84. Therefore, we would conclude that there is a statistically significant relationship between gender and preferred spectator sport. You can also determine statistical significance by comparing alpha and the P-value. If the P-value is less than alpha, then the result is statistically significant.

| Section 13.1 | **Random Selection Using the Random Number Generation Tool** |

Researchers often need to select a random sample of people to participate in a research project. Excel's Random Number Generation tool provides an easy way to do this. The Random Number Generation tool samples with replacement, so it is possible that the same participant could be selected more than once.

Sample Research Problem

A customer service department wanted to survey a random sample of five customers who had phoned for assistance during the past week. Each of the 20 customers who had phoned was assigned a customer number from 1 to 20. The probability of randomly selecting any one of these customers was 1/20 or .05.

1. Open worksheet "Ch13_Customer" on the Web site, or enter the labels, customer numbers, and the probabilities in an Excel worksheet as shown below.

	A	B	C
1	Customer	Probability	Sample
2	1	0.05	
3	2	0.05	
4	3	0.05	
5	4	0.05	
6	5	0.05	
7	6	0.05	
8	7	0.05	
9	8	0.05	
10	9	0.05	
11	10	0.05	
12	11	0.05	
13	12	0.05	
14	13	0.05	
15	14	0.05	
16	15	0.05	
17	16	0.05	
18	17	0.05	
19	18	0.05	
20	19	0.05	
21	20	0.05	

2. Click the **Data** tab near the top of the screen and select **Data Analysis**.

If Data Analysis does not appear as a choice in the Data ribbon, you will need to load the Microsoft Excel ToolPak add-in. Follow the procedure on page 9.

3. Select **Random Number Generation** in the Data Analysis dialog box. Click **OK**.

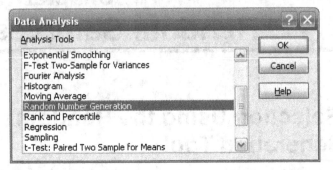

4. Complete the Random Number Generation dialog box as shown below. Click **OK**.

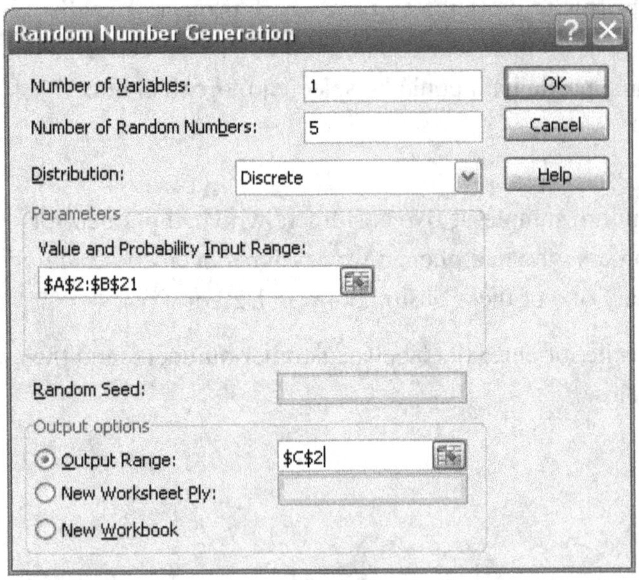

5. The random sample shown below contains a repetition of customer number 18. If repetitions occur in your sample, repeat the procedure to generate more numbers. To do this, first select **Data Analysis** in the Data ribbon.

	A	B	C
1	Customer	Probability	Sample
2	1	0.05	8
3	2	0.05	3
4	3	0.05	12
5	4	0.05	18
6	5	0.05	18

6. Select **Random Number Generation** in the Data Analysis dialog box. Click **OK**.

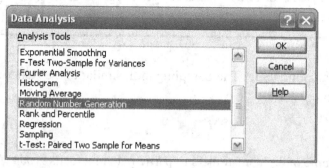

7. Complete the Random Number Generation dialog box as shown below. Be sure to use cell C7 for the output range. Click **OK**.

The list now enables you to select five different customers for the survey project. Those customers are numbers 8, 3, 12, 18, and 13. If any of these customers decline to participate, you can continue with customer number 20. Because the numbers were generated randomly, your list will likely have different numbers.

	A	B	C
1	Customer	Probability	Sample
2	1	0.05	8
3	2	0.05	3
4	3	0.05	12
5	4	0.05	18
6	5	0.05	18
7	6	0.05	13
8	7	0.05	20
9	8	0.05	20
10	9	0.05	12
11	10	0.05	5

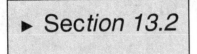

Random Selection Using the Sampling Tool

Samples can be selected randomly using Excel's Sampling tool.   The Sampling tool, similar to the Random Number Generation tool, samples with replacement.

Sample Research Problem

A market researcher wanted to select a random sample of eight customers to participate in a focus group. The researcher was selecting from a pool of 15 customers who indicated their willingness to participate.

1. Open a new Excel worksheet and enter the label **Customer** and the numbers 1 through 15 as shown below.

	A
1	Customer
2	1
3	2
4	3
5	4
6	5
7	6
8	7
9	8
10	9
11	10
12	11
13	12
14	13
15	14
16	15

2. Click the **Data** tab near the top of the screen and select **Data Analysis**.

If Data Analysis does not appear as a choice in the Data ribbon, you will need to load the Microsoft Excel ToolPak add-in. Follow the procedure on page 9.

3. Select **Sampling** in the Data Analysis dialog box. Click **OK**.

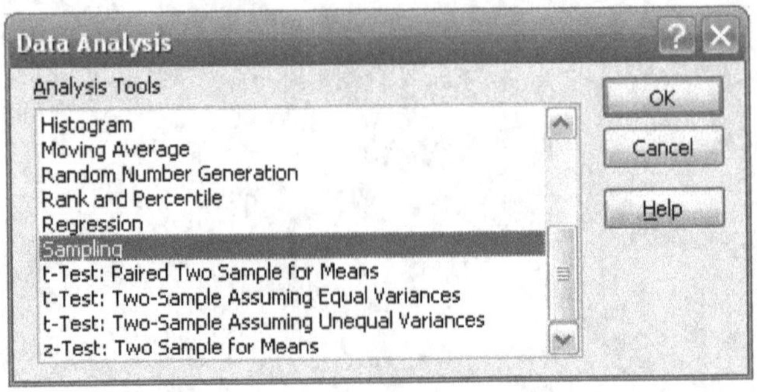

Data Analysis

Analysis Tools

Histogram
Moving Average
Random Number Generation
Rank and Percentile
Regression
Sampling
t-Test: Paired Two Sample for Means
t-Test: Two-Sample Assuming Equal Variances
t-Test: Two-Sample Assuming Unequal Variances
z-Test: Two Sample for Means

OK
Cancel
Help

4. Complete the Sampling dialog box as shown below. Click **OK**.

```
┌─ Sampling ─────────────────────────── ? X ─┐
│  ┌─ Input ──────────────────────────┐   ┌──────────┐  │
│  │  Input Range:      $A$1:$A$16  ▦ │   │    OK    │  │
│  │  ☑ Labels                        │   └──────────┘  │
│  └──────────────────────────────────┘   ┌──────────┐  │
│  ┌─ Sampling Method ────────────────┐   │  Cancel  │  │
│  │  ○ Periodic                      │   └──────────┘  │
│  │     Period:     [          ]     │   ┌──────────┐  │
│  │                                  │   │   Help   │  │
│  │  ◉ Random                        │   └──────────┘  │
│  │     Number of Samples:  [ 8  ]   │                 │
│  └──────────────────────────────────┘                │
│  ┌─ Output options ─────────────────┐                 │
│  │  ◉ Output Range:   [ $B$1| ]  ▦  │                 │
│  │  ○ New Worksheet Ply: [       ]  │                 │
│  │  ○ New Workbook                  │                 │
│  └──────────────────────────────────┘                │
└────────────────────────────────────────────┘
```

5. Because these numbers were generated randomly, it is not likely that your output will be exactly the same as mine. There were two repetitions in the output. Numbers 11 and 10 appear twice. So, we need to repeat the procedure. Click the **Data** tab near the top of the screen and select **Data Analysis**.

	A	B
1	Customer	11
2	1	3
3	2	11
4	3	4
5	4	14
6	5	10
7	6	10
8	7	7

6. Select **Sampling** in the Data Analysis dialog box. Click **OK**.

7. Complete the Sampling dialog box as shown below. Be sure to use cell B9 for the output range. Click **OK**.

```
┌─ Sampling ─────────────────────────── ? X ─┐
│  ┌─ Input ──────────────────────────┐   ┌──────────┐  │
│  │  Input Range:      $A$1:$A$16  ▦ │   │    OK    │  │
│  │  ☑ Labels                        │   └──────────┘  │
│  └──────────────────────────────────┘   ┌──────────┐  │
│  ┌─ Sampling Method ────────────────┐   │  Cancel  │  │
│  │  ○ Periodic                      │   └──────────┘  │
│  │     Period:     [          ]     │   ┌──────────┐  │
│  │                                  │   │   Help   │  │
│  │  ◉ Random                        │   └──────────┘  │
│  │     Number of Samples:  [ 8  ]   │                 │
│  └──────────────────────────────────┘                │
│  ┌─ Output options ─────────────────┐                 │
│  │  ◉ Output Range:   [ $B$9 ]  ▦   │                 │
│  │  ○ New Worksheet Ply: [       ]  │                 │
│  │  ○ New Workbook                  │                 │
│  └──────────────────────────────────┘                │
└────────────────────────────────────────────┘
```

The focus group will include customers who were assigned the numbers 11, 3, 4, 14, 10, 7, 1, and 6.

	A	B
1	Customer	11
2	1	3
3	2	11
4	3	4
5	4	14
6	5	10
7	6	10
8	7	7
9	8	1
10	9	4
11	10	11
12	11	3
13	12	11
14	13	10
15	14	6
16	15	7

▶ Section 13.3   Random Selection Using the RANDBETWEEN Function

The RANDBETWEEN function returns a randomly selected number between the numbers specified by the researcher. This function is especially useful if the researcher is selecting from a very large range of numbers.

Sample Research Problem

An energy conservation firm in a large city wanted to find out if home buyers took energy-saving features into consideration when they bought their house. During the last year, the total number of home sales in the city was 2,000. The firm wanted to randomly select 20 of these home buyers to receive a free energy audit and to participate in a personal interview.

1. Each home buyer was given an identification number, from 1 to 2,000. We will start by selecting one home buyer. Click in cell **A1** where the buyer's number will be placed.

The RANDBETWEEN function is in the ToolPak add-in. The function will return the #NAME? error if the Analysis ToolPak is not available. Instructions for loading the Microsoft Excel ToolPak add-in are given on page 9.

2. Click the **Formulas** tab near the top of the screen and select **Insert Function**.

3.   Select the **All** category. Then scroll down to select the **RANDBETWEEN** function. Click **OK**.

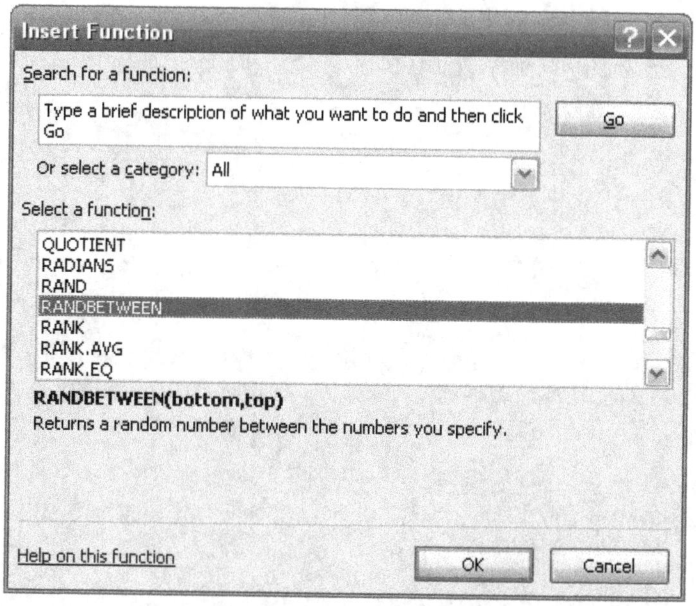

Insert Function   `? X`

Search for a function:

Type a brief description of what you want to do and then click Go   Go

Or select a category: All

Select a function:

QUOTIENT
RADIANS
RAND
RANDBETWEEN
RANK
RANK.AVG
RANK.EQ

RANDBETWEEN(bottom,top)
Returns a random number between the numbers you specify.

Help on this function   OK   Cancel

4.   Type **1** in the Bottom window. Type **2000** in the Top window. Click **OK**.

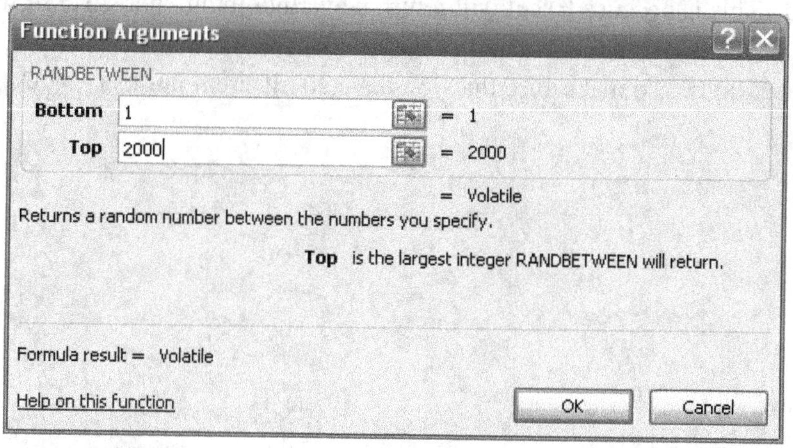

Function Arguments   `? X`

RANDBETWEEN

Bottom 1   = 1
Top 2000   = 2000

= Volatile
Returns a random number between the numbers you specify.

Top   is the largest integer RANDBETWEEN will return.

Formula result = Volatile

Help on this function   OK   Cancel

5.   The function returns 573. So, we would ask the buyer with identification number 573 to participate in the energy audit and personal interview. (Because the number was generated randomly, it is not likely that your output will be the same.)

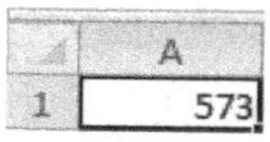

6.   Let's say that the firm wants a total of 20 randomly selected home buyers. To generate a random sample of 20 numbers between 1 and 2,000, just copy the function in cell **A1** to cells A2 through A20.

	A
1	301
2	435
3	46
4	822
5	18
6	1031
7	121
8	1240
9	993
10	1551
11	928
12	1137
13	160
14	332
15	1916
16	1914
17	894
18	154
19	1886
20	1750

Note that the number in cell A1 changed. This is because Excel will return new random numbers every time the worksheet is calculated. The RANDBETWEEN function samples with replacement. So, it is possible that your list of 20 numbers will have repetitions. To make sure that you have 20 different numbers, you may want to generate a sample of 25 to 30.

Index

A
ABS function 135-137, 149
Absolute references 34
Activate a cell 14
Active cell 14
Add-Ins 9
Addresses 14
Adjusted R square 209, 213, 219, 223, 226
Analysis of variance 171
Analysis of variance compared to dummy coding in regression 223
Analysis Toolpak 9-11
ANOVA: Single factor 172-175
ANOVA: Two-factor with replication 181-185
ANOVA: Two-factor without replication 176
AutoSum 37, 42
AVEDEV function 87, 88-89, 91
AVERAGE function 87, 89-90, 92, 132-133, 141
Axis titles 53, 65, 72, 195-196

B
Bar chart 57-58
BINOM.DIST function 104-107
Binomial distribution 103-104

C
Chart title 53, 195, 229
CHISQ.DIST.RT function 129-130
CHISQ.INV.RT function 130, 254
Chi-square distribution 115, 129
Chi-square test of independence 249-255
CHISQ.TEST function 254
Click 3
Column width 22
Confidence interval 139-140, 152-153
Confidence level 139, 153
Continuous probability distributions 115-130

Copying a formula 41-42, 46
Copying an entire worksheet 20
Copying information 19
CORREL function 190-192
Correlation 189-203
Correlation analysis tool 192-194
Correlation matrices 192
Count in Descriptive Statistics output 77, 79, 81
COUNT function 37-39, 87, 133-134, 143-144
Cross tabulations 233-255
Cross tabulations using the Pivot Table 233-249
Currency format 24
Curvilinear regression 224-232
Cut 19-20

D
Data analysis tools 77-87, 156-170, 172-188, 192-194, 206-228, 257-262
Data files for sample problems on a Web site 5
Data labels 53
Data ribbon 4
Data table in a chart 66
Decimal points 23
Deleting rows and columns 21-22
Descriptive statistics 77-102
Design ribbon 52-53, 58
Deviation score 36, 40-41, 45-46
Dialog boxes 5
Discrete probability distributions 103-115
Display regression equation on chart 231
Display R-squared on chart 198, 231
Double-click 2
Drag 3, 4
Dummy coding 215-224

E
Editing information 18-22

Equations 32-34
Exiting Excel 3
Exponentiation in formulas 31
F
F distribution 126-128
F.DIST.RT function 126-127
File name 6
F.INV.RT function 126-128
F-test 171, 175, 180, 185
F-test for two sample variances 185
Fill handle 15-16
Filling a series 16-18
Filling adjacent cells 15-16
Footer 6, 8
Format data series 53, 62, 70
Formatting numbers 22-25
Formula bar 14
Formulas 31-48
Formulas ribbon 4
Formulas to calculate expected cell frequencies 252-254
Frequency distribution for grouped data 59-67
Frequency distribution for ungrouped data 68-74
Frequency distribution of a qualitative variable 55-58
Frequency distribution of a quantitative variable 49-54
FREQUENCY function 74-76
Functions for a chi-square test of independence 254
Functions for descriptive statistics 87-92

G
Gridlines 8-9
Gridlines in a graph 73, 197

H
Header 6-8
Histogram analysis tool 49, 54, 59-74
Histogram for grouped data 59-67
Histogram for ungrouped data 68-74
Histogram of a quantitative variable 49-54

Home ribbon 3
Hypergeometric distribution 108
HYPGEOM.DIST function 108-111

I
IF statement 216
Insert Function 4
Insert ribbon 3
Inserting or deleting rows and columns 21-22
Intercept of a regression line 205, 209, 214, 219, 223, 227

K
KURT function 87
Kurtosis 77, 79, 81

L
Largest value in a distribution 79, 81, 86
LARGE function 87
Layout ribbon 53, 66, 195-196
Layout of worksheets 3-4
Legend 53, 58, 65-66, 72, 195, 230
Line fit plots 215

M
Margins 6-7
Mathematical operators in formulas 31
MAX function 87
Maximum 77, 79, 81, 92
Mean 36-40, 77, 79, 87, 89, 96
Median 77, 79, 80, 87
MEDIAN function 87
MIN function 87
Minimum 77, 79, 81, 87, 92
Missing data 87, 247-249
Mode 77, 79, 80, 87, 88
MODE.MULT function 87
MODE.SNGL function 88
Mouse 2-3
Moving an entire worksheet 20
Moving information 19
Multiple regression 210-215

N
Name box 14

Normal distribution 115-122
NORM.DIST function 115
NORM.INV function 116
NORM.S.DIST function 116, 117-119
NORM.S.INV function 113, 116, 119-122, 137
Numeric information 15

O

One-sample t-test 140-153
One-sample z-test 131-140
One-way between-groups ANOVA 171-175
One-way repeated measures ANOVA 175-179
Opening a brand new worksheet 13
Opening a file 13
Order of mathematical operations in formulas 31-32

P

Page Layout ribbon 4
Page setup for printing 4, 6-9
Paired-samples t-test 163-167
Paste 19
Pearson correlation coefficient 189-194, 202
Percentage format 24-25
Pivot Chart 49, 52-54, 57-58
Pivot Table 45, 49-52, 55-57, 233-249
Pivot Table descriptive statistics 77, 79, 92-98
Pivot Table Field List 50
Pivot Table for observed cell frequencies 250-252
Poisson distribution 111-115
POISSON.DIST function 111-115
Polynomial trendline 231
Population standard deviation 82-83
Population variance 82-84
Printing 6-9
Probability distributions 103-130

Q

QUARTILE.INC function 88

R

R square 209, 213, 219, 223, 226
RANDBETWEEN function 262-264
Random number generation tool 257-259
Random samples 257-264
Random selection 257-264
Range in descriptive statistics output 77, 79, 81
Range of cells 14
RANK.EQ function 200-202
Recoding 25-27
Regression 205-232
Regression analysis tool 206-228
Regression coefficients 209, 212, 214, 219, 223, 227
Relative references 34, 35
Residuals 210, 212, 214, 217, 220, 221, 224, 227
Residual plots 215
Right-click 3

S

Sampling tool 260-262
Saving files 6
Scatterplot 194-199
Scatterplot of a curvilinear relation 228-232
Scroll bars 4
Series 15, 16-18
SKEW function 88
Skewness 77, 79, 81
Slope of a regression line 205, 209, 210
Smallest value in a distribution 79, 81, 86
SMALL function 88
Sorting 28-29
Spearman rank correlation coefficient 199-203
Squared deviation scores 36, 41-42
SQRT function 44-45
Standard deviation 44-45
Standard deviation of a population 82-83, 88

Standard error of the mean 79, 80, 81, 84-85, 131, 134, 138, 139, 140, 144, 151, 152
STANDARDIZE function 116-117
Starting Excel 3
STDEV.P function 82-83, 87
STDEV.S function 87, 142-143
Sum in Descriptive Statistics output 77, 79
Sum in Pivot Table output 92
SUM function 36-37, 42

T
t distribution 122-126
T.DIST function 145-146
T.DIST.RT function 122, 123-124
T.DIST.2T
Text information 13, 60
T.INV function 121, 147, 150
T.INV.2T function 122-123
Trendline 197-198, 228, 230-231,
t^*-test 159-163
t-test for one sample 140-153
t-test for paired samples 163-167
t-test for two independent samples 155-163
t-test: Paired two sample for means 164
t-test: Two-sample assuming equal variances 157
t-test: Two-sample assuming unequal variances 161
Two-variable regression 205-210
Two-way between-groups ANOVA 180-185

V
Variance 36, 42-44, 77, 79, 81,
Variance of a population 42-44, 83-84, 85
Variance of a sample 88
VAR.P function 83-84, 88
VAR.S function 88
Versions of Excel 1
Versions of Windows 2
VLOOKUP function 26-27

W
Web site for data files for sample problems 5
Weighting to adjust for survey nonresponse 99-102
Welch-Aspin procedure 159-160
Worksheet area 4
Worksheet tabs 4

Z
z-scores 36, 45-47
z-scores using STANDARDIZE function 116-117
z-test for one-sample mean 131-140
z-test for two independent samples 167-170
z-test: Two sample for means 168